SpringerBriefs in Earth System Sciences

Series Editors
Kevin Hamilton
Gerrit Lohmann
Lawrence A. Mysak

For further volumes:
http://www.springer.com/series/10032

Sophie Valcke · René Redler
Reinhard Budich

Earth System Modelling – Volume 3

Coupling Software and Strategies

 Springer

Dr. Sophie Valcke
CERFACS, Av. Coriolis 42
31057 Toulouse Cedex 01
France
e-mail: valcke@cerfacs.fr

Dr. René Redler
Max-Planck-Institut für Meteorologie
Bundesstraße 53
20146 Hamburg
Germany
e-mail: rene.redler@zmaw.de

Reinhard Budich
Max-Planck-Institut für Meteorologie
Bundesstraße 53
20146 Hamburg
Germany
e-mail: reinhard.budich@zmaw.de

ISSN 2191-589X   e-ISSN 2191-5903
ISBN 978-3-642-23359-3   e-ISBN 978-3-642-23360-9
DOI 10.1007/978-3-642-23360-9
Springer Heidelberg Dordrecht London New York

Library of Congress Control Number: 2011938153

Printed on acid-free paper

Springer is part of Springer Science+Business Media (www.springer.com)

To all people who work hard every day to make the world a better place

Preface

Climate modelling in former times mostly covered the physical processes in the atmosphere. Nowadays, there is a general agreement that not only physical, but also chemical, biological and, in the near future, economical and sociological—the so-called anthropogenic—processes have to be taken into account on the way towards comprehensive Earth system models. Furthermore these models include the oceans, the land surfaces and, so far to a lesser extent, the Earth's mantle. Between all these components feedback processes have to be described and simulated.

Today, a hierarchy of models exist for Earth system modelling. The spectrum reaches from conceptual models—back of the envelope calculations—over box-, process- or column-models, further to Earth system models of intermediate complexity and finally to comprehensive global circulation models of high resolution in space and time. Since the underlying mathematical equations in most cases do not have an analytical solution, they have to be solved numerically. This is only possible by applying sophisticated software tools, which increase in complexity from the simple to the more comprehensive models.

With this series of briefs on "Earth System Modelling" at hand we focus on Earth system models of high complexity. These models need to be designed, assembled, executed, evaluated, and described, both in the processes they depict as well as in the results the experiments carried out with them produce. These models are conceptually assembled in a hierarchy of sub models, where process models are linked together to form one component of the Earth system (Atmosphere, Ocean, ...), and these components are then coupled together to Earth system models in different levels of completeness. The software packages of many process models comprise a few to many thousand lines of code, which results in a high complexity of the task to develop, optimise, maintain and apply these packages, when assembled to more or less complete Earth system models.

Running these models is an expensive business. Due to their complexity and the requirements w.r.t. the ratios of resolution versus extent in time and space, most of these models can only be executed on high performance computers, commonly called supercomputers. Even on todays supercomputers typical model experiments take months to conclude. This makes it highly attractive to increase the efficiency of the codes. On the other hand the lifetime of the codes excesses the typical lifetime of computing systems, including the specifics of their architecture, roughly by a factor of 3. This means that the codes need not only be portable, but also constantly adapted to emerging computing technology. Whereas in former times computing power of single processors—and such of clustered computers— was resulting mainly from increasing clock speeds of the CPUs, todays increases are only exploitable when the application programmer can make best use of the increasing parallelity off-core, on-core and in threads per core. This puts additional complexity in areas like IO performance, communication between cores or load balancing to the assignment at hand.

All these requirements put high demands on the programmers to apply software development techniques to the code making it readable, flexible, well structured, portable and reusable, but most of all capable in terms of performance. Fortunately these requirements match very well an observation from many research centres: due to the typical structure of the staff of the research centres code development oftentimes has to be done by scientific experts, who typically are not computing or software development experts. So code they deliver deserves a certain quality control to assure fulfilment of the requirements mentioned above. This quality assurance has to be carried out by staff with profound knowledge and experience in scientific software development and a mixed background from computing and science.

Since such experts are rare, an approach to ensure high code quality is the introduction of common software infrastructures or frameworks. These entities attempt to deal with the problem by providing certain standards in terms of coding and interfaces, data formats and source management structures etc., that enable the code developers as much as the experimenters to deal with their Earth system models in a well acquainted, efficient way. The frameworks foster the exchange of codes between research institutions, the model inter-comparison projects so valuable for model development, and the flexibility of the scientists when moving from one institution to another, which is commonplace behaviour these days.

With an increasing awareness about the complexity of these various aspects scientific programming has emerged as a rather new discipline in the field of Earth system modelling. Coevally new journals are launched providing platforms to exchange new ideas and concepts in this field. Up to now we are not aware of any text book addressing this field, tailored to the specific problems the researcher is confronted with. To start a first initiative in this direction, we have compiled a series of six volumes each dedicated to a specific topic the researcher is confronted with when approaching Earth System Modelling:

1. Recent developments and projects
2. Algorithms, code infrastructure and optimisation
3. Coupling software and strategies
4. IO and postprocessing
5. Tools for configuring, building and running models
6. ESM data archives in the times of the grid

This series aims at bridging the gap between IT solutions and Earth system science. The topics covered provide insight into state-of-the-art software solutions and in particular address coupling software and strategies in regional and global models, coupling infrastructure and data management, strategies and tools for pre- and post-processing, and techniques to improve the model performance.

Volume 1 familiarizes the reader with the general frameworks and different approaches for assembling Earth system models. Volume 2 highlights major aspects of design issues that are related to the software development, its maintenance and performance. Volume 3 provides an overview and introduction into the major coupling software products developed and used in the climate modelling community. The different technical attempts or possibilities are discussed from the software point of view to solve the coupled problem. Once the coupled model is running data are produced and postprocessed (Volume 4). The whole process of creating the software, running the model and processing the output is assembled into a workflow (Volume 5). Volume 6 describes coordinated approaches to archive and retrieve data.

Hamburg, June 2011   Reinhard Budich
                                                                                            René Redler

Acknowledgments

We would like to thank all authors for their tenacity in the writing of this Volume, and Michel Desgagné and Anthony Thevenin for their very constructive review.

Contents

Contributors

V. Balaji Princeton University, New Jersey, USA, e-mail: balaji@princeton.edu

Cecelia DeLuca NOAA/CIRES, Boulder, CO, USA, e-mail: cecelia.deluca@noaa.gov

Rupert Ford University of Manchester, Manchester, UK

Robert Jacob Argonne National Laboratory, Argonne, USA, e-mail: jacob@mnsc.anl.gov

Jay Larson Research School of Computer Science, The Australian National University, Canberra, Australia, e-mail: larson@mnsc.anl.gov

Ciaron Linstead Potsdam Institute for Climate Impact Research, Potsdam, Germany, e-mail: linstead@pik-potsdam.de

René Redler Max-Planck-Institut für Meteorologie, Bundesstraße 53, 20146, Hamburg, Germany, e-mail: rene.redler@zmaw.de

Graham Riley University of Manchester, Manchester, UK, e-mail: graham.riley@manchester.ac.uk

Gerhard Theurich Science Applications International Corporation, McLean, VA, USA,

Sophie Valcke CERFACS, Av. Coriolis 42, 31057, Toulouse Cedex 01, France, e-mail: sophie.valcke@cerfacs.fr

Chapter 1
Introduction

Sophie Valcke

Coupling numerical codes is certainly not a new preoccupation in the climate research community and in other research fields such as electromagnetism or computational fluid dynamics. The climate models in particular are almost never written "from scratch": they consist of numerical models of the subsystems of the Earth's climate joined together. The granularity of these components is typically relatively coarse (e.g. the atmosphere, the ocean, the land surface, etc.) but it can, in some cases, be finer (e.g. the atmospheric radiation, the ocean vertical mixing, etc.). These components are complex applications themselves, gathering hours of development by scientists with highly specialized skills and knowledge.

Thus the creation of a coupled system involves coding an additional software interface to the separately developed components containing the science of climate modelling. The main functions shared by the different coupling software currently used by the climate modelling groups are to transfer gridded fields between component models and to interpolate these fields from the source grid to the target one in an efficient, consistent and extendible manner.

The design of the coupling software can follow different basic approaches. On one hand, the design can aim at a minimal level of interference in the existing codes ensuring that they will run as separate executables with main characteristics unchanged with respect to the standalone mode. This approach, while in some cases less efficient, is probably the best trade-off that can be chosen when strict external coding rules are not likely to be observed by the groups developing the components. On the other hand, when a deeper level of modifications in the existing codes is acceptable, the principle is to build a hierarchical coupled application integrating the different parts of the original codes; this approach leads in general to a more efficient coupled application and is therefore probably most recommended in a more controlled development environment.

S. Valcke (✉)
CERFACS, Av. Coriolis 42, 31057 Toulouse Cedex 01, France
e-mail: sophie.valcke@cerfacs.fr

S. Valcke et al., *Earth System Modelling – Volume 3*,
SpringerBriefs in Earth System Sciences, DOI: 10.1007/978-3-642-23360-9_1,

This volume presents in more details the major coupling softwares developed and used in the international climate modelling community, each one implementing to a lesser or greater extent one of these approaches. The simple TDT library (see Chap. 2) provides a light means of transferring data in a platform and language independent way. MCT (see Chap. 3), used in particular at the National Center for Atmospheric Research in the USA for the Community Climate System Model, is a lower level library that provides datatypes and methods to create coupler entities or parallel coupled models. The OASIS coupler (see Chap. 4) widely used in the diverse European climate modelling community to couple independently developed state-of-the-art climate component models is a good example of the former design approach. The latter approach is adopted for example by FMS (see Chap. 5) developed by the Geophysical Fluid Dynamics Laboratory (GFDL) which offers a comprehensive programming model and toolkit for the construction of coupled climate models and by the ESMF software infrastructure (see Chap. 6) developed in the USA which follows a component based design to build climate applications as an hierarchy of nested components. Finally, BFG (see Chap. 7) developed by the University of Manchester differs from the other coupling approaches described here as it is designed to allow the user to choose a coupling technology (e.g. OASIS or ESMF) and to generate related wrapper code around the user science code, thereby creating a coupled application based on the chosen coupling technology.

Chapter 2
TDT: A Library for Typed Data Transfer

Ciaron Linstead

2.1 Introduction

Many solutions exist for the problem of model coupling such as a fully integrated coupler like OASIS, geared towards large model coupling problems, or more basic communication layers like the Common Object Request Broker Architecture (CORBA), which can nevertheless require a large investment of effort to implement.

Model coupling in climate research, or any other field, comprises several levels. At it is lowest level, which we address here, model coupling is a technical exercise in transferring data between two compatible programs. The compatibility of these programs requires, for example, knowledge of the physical processes being coupled and the numerical schemes being implemented to model those processes.

The Typed Data Transfer (TDT) library addresses the technical coupling of modules and provides a means of transferring data between programs in a platform and programming language independent way. This is achieved by means of a library of functions which can be used in C, Fortran, Python and Java programs, or any programming language which can import foreign functions from a C library.[1] Heterogeneous coupling, i.e. transfer of data between remote machines, is also supported.

The TDT will transfer primitive data types (integer, float, char, etc.) as well as complex data structures (arrays and matrices, C structures) with one function call.

Non-TDT programs that use the same underlying transfer protocols can also communicate with TDT programs. TDT's output is simply byte-streams which can

---

[1] All the examples given here will be in C. Examples of the TDT in use in Fortran and Python are available in the standard TDT distribution.

C. Linstead (✉)
Potsdam Institute for Climate Impact Research,
Potsdam Germany
e-mail: linstead@pik-potsdam.de

S. Valcke et al., *Earth System Modelling – Volume 3*,
SpringerBriefs in Earth System Sciences, DOI: 10.1007/978-3-642-23360-9_2,
© The Author(s) 2012

be decoded by receiving programs, though at the expense of replicating functionality which already exists in the TDT library.

2.2 Architectural Overview

The TDT has been designed and implemented with several principles in mind:

Ease of Use

The intention from the outset was to design a library which would read or write data with function calls no more complicated than a simple "print" statement. The investment of time and effort by a programmer in simply trying out the TDT is significantly lower than for other, more complex, data transfer or model coupling mechanisms.

Portability

With the desire to share scientific code modules and the trend towards distributed model coupling, it is important that modules be designed to operate on many operating systems and interface with as many programming languages with as little modification as possible. In achieving this, it was decided at an early stage to write the core functionality in C (following the Portable Operating System Interface (POSIX) standard) and use the American National Standards Institute (ANSI) standard for data types. Most major programming languages implement a foreign function interface for C libraries, making the task of using the TDT in, for example, Fortran a matter of writing call-through functions.

Efficiency

The efficiency of data transfers is a major consideration in designing a model coupling scheme, particularly with complex climate systems models. An inefficient mechanism for moving data would be an immediate barrier to its adoption. Thus, the TDT does not perform any marshalling or buffering of data, and no data conversions except byte swapping where necessary. Contiguous blocks of similar data types are sent with one transfer, regardless of whether they belong to the same data structure in the communicating programs. Tests of transfers over raw TCP sockets have shown that there is little difference in speed between TDT transfers and transfers over hand-crafted socket connections.

Flexibility

The TDT is designed such that extensions to its capabilities are relatively straightforward. New data types can be added, as can new fundamental data transfer protocols. The code has been released under the GNU Public License (GPL) so the programmer has complete freedom to extend the library.

Using the TDT from a new programming language is straightforward when the language provides a foreign function interface, although the TDT has been successfully employed in coupling programs written in languages where no foreign function interface exists (e.g. the General Algebraic Modeling System (GAMS)).

Usefulness

In designing the TDT for ease of use another property emerges, that of usefulness. Fitting the "Unix philosophy"—do one thing well—it can be applied to many fields with equal effectiveness and does not require specialist knowledge of a particular subject area in order to be useful.

2.3 Benefits and Limitations of TDT

Transferring data between programs requires that both programs implement code for reading and writing this data. The implementation of this code depends on several things:

- data types
- size of data (e.g. length of arrays)
- the transfer mechanism, or protocol
- byte order, or endianness

Implemented in the traditional way, each program would contain hard-coded information about the data it sends and receives. The TDT aims to minimise this by keeping as much of this information as possible in external descriptions of the data transfer.

Data Types

The implementation of the data transfer code needs to take account the types of the data being transferred. Different architectures may represent the same data type with a different number of bytes and the programmer would normally be expected to know at compile-time how to translate these data types. This adds complexity to the code, and reduces the flexibility with which the coupling can use different data types.

Size of Data

If the size of a data structure being transferred is altered, several changes need to be made to each program. First, the internal structure for storing this data must be changed. Then, the code which handles the sending or receiving of the data needs to be altered to reflect the new size. The TDT library removes the necessity for this second step, by representing the data structure externally in a data description file. This makes program modifications easier and less error prone.

Transfer Mechanism

The underlying protocols supported by the TDT for the transfer of data are: Unix sockets, intermediate files (either on disk or in shared memory), or higher level methods such as the Message Passing Interface (MPI) for transfer between parallel processors. Usually, switching between these protocols requires major modifications to code, with the attendant risk of introducing bugs; with the TDT, the choice of protocol is made by changing a parameter in the modules configuration file. The choice of protocol can thus be changed at various stages in the development process, depending on whether, for example, modules must be distributed (using network sockets) or must communicate very fast (using shared memory).

Byte Swapping

When data is represented in multiple bytes, the order of those bytes in memory is dependent on the architecture of the machine. This is called endianness and has two main alternatives, big endian and little endian. Transferring data between machines with different endianness requires that bytes be swapped at one end of the transfer. The TDT library handles byte swapping internally and is transparent to the developer.

Limitations of TDT

The TDT handles the low-level software infrastructure of data transfers between programs. The semantics of those transfers must make sense at the physical, mathematical and numerical levels to have the coupled model give meaningful results. This is not checked by the TDT, although the data descriptions which the programmer must write can act as basic documentation of the technical interfaces between modules.

The TDT itself performs no data processing, such as interpolation, but one could envisage a scenario where an interpolation "module" is plugged into the transfer using TDT.

Synchronising modules must also be checked manually. Each "read" must be matched by a corresponding "write".

2.4 Model Coupling with TDT

A model coupled with TDT has several elements:

- the modules being coupled
- the data being transferred
- a configuration file per module (XML)
- the description of the data being transferred (XML)
- communication mechanism, e.g. files, sockets etc.

The coupling is a transfer of data between two programs at some point in their execution. This transfer is carried out directly between the components themselves—the TDT does not buffer or marshal the data which would have a detrimental effect on the performance of the transfer.

The timing of the transfer is determined by where in the module code the "read" and "write" function calls are inserted. The matching of read and write operations must be done by the programmer to avoid lost data or deadlock situations.

This transfer occurs over a defined channel which has two principal characteristics: the data it will carry and the type of the underlying transfer mechanism (e.g. sockets, files, shared memory). These two elements are described in two configuration files, the data description file (datadesc) and the configuration file (config).

These files are in eXtensible Markup Language (XML) format, a standard language for metadata markup which has many freely available tools for processing and display.

Each end of the transfer (i.e. the reader and the writer) reads the data description file (or has its own copy of the file in the case of distributed models) which the TDT parses on initialisation. The datadesc can contain descriptions of all data being transferred on a particular channel (multiple datadesc files for multiple channels) or it can describe all data on all channels (one datadesc file for multiple channels).

The datadesc contains declarations of variables, their types and sizes if necessary. A full description of the syntax is given in Sect. 2.5.1.

Calls to underlying system-dependent functions for sending and receiving data are dispatched from the TDT read and write functions (tdt_read() and tdt_write()) according to this datadesc. This is carried out recursively when necessary, for example when transferring multidimensional arrays or C structures.

Each module in the coupled system has its own configuration file. This file defines which input and output channels the program uses, the transfer protocols for these channels, the name of the datadesc file which defines this channels data and any connection-specific information for the channel, for example port numbers and hostnames for socket communications. A description of the syntax used in the config file is given in Sect. 2.5.2.

2.5 TDT in Practice

To use the TDT the following steps, in addition to the normal programming of the module, must be followed:

- create data description and configuration files
- include TDT header file
- declare variables required by TDT
- read the configuration file
- open the communication channel
- perform the read or write
- close the channel

  Other, optional, function calls are:

- *resize an array*: this is useful for dynamic arrays whose size is not known when the data description is written
- *send and receive the address of a TDT program*: this is useful when location of one endpoint of the communication is not known in advance, for instance when writing a controller for communication between several modules.

  This section includes a simple example of a two-way data transfer between two programs. This example is included in the TDT source code distribution.

2.5.1 The Data Description File

The data description file contains one or more declarations which describe the types of data which will be transferred.

  The general format of the datadesc is:

```
< data_desc >

    <decl name="string">
        datatype
    </decl>
      .
      .
      .
</data_desc>
```

where "datatype" is one of the primitive datatypes:

- `int`
- `float`
- `double`
- `char`

or a complex datatype:

- `< struct>`
  ```
  <decl name="string">
  datatype
  </decl>
  .

  .
  </struct>
  ```

- `< array size="integer">`
  ```
    datatype
    < /array>
  ```

The "name" attribute of the `<decl>` tag uniquely identifies the data item to the TDT. The calls to `tdt_read()` and `tdt_write()` in the modules use both this name and the variable name to specify to the TDT the data being transferred. In practice, making the name attribute the same as the variable name in the program contributes to good documentation of the interface between the coupled modules. Note that this grammar allows nested arrays. These are handled recursively by the TDT and are transferred with a single function call.

2.5.2 The Configuration File

The TDT configuration files define the channels particular modules will communicate on. The general format is as follows:

```
< program name="string">

    < channel name="string"
              mode=in | out
              type=socket | file
              filename="string"
              host="string"
              port="integer"
              datadesc="string">
    < /channel>

< /program>
```

The <channel> tag has several attributes. These are:

- name: the name of the channel. This is used when opening the channel
- mode: in or out, the direction this data flows
- type: socket, file or mpi, the communication protocol to use
- filename: if type is file, the filename to read from or write to.
- host: if type is socket, the hostname to read from or write to.
- port: if type is socket, the port number to read from or write to.
- datadesc: the name of the file containing the description of the data this channel will transfer.

The configuration file will contain descriptions of all the channels a particular program uses throughout the course of its execution.

2.5.3 Implementation in the Original Codes

Include TDT Header File

Simply add the include directive #include "tdt.h" at the start of the module code.

Declare Variables

Two variables are required for storing TDT specific information, one of type TDTConfig and the other TDTState. One TDTConfig is required per configuration file being read (generally only one) and one TDTState per channel.

Read Configuration File

This step is carried out once per configuration file (i.e. generally once during execution of a program). It reads and parses the configuration file and stores the resulting data structure in the TDTConfig variable previously declared:
tc = tdt_configure ("configuration-filename").

Open Communication Channel

A call to the tdt_open() function prepares the communication channel for read and write operations. A typical call to this function looks like this:
    ts = tdt_open (tc, "channel-name");
where ts is the TDTState variable, and tc the TDTConfig variable. The channel identified by channel-name must be described in the configuration file

(i.e. "`configuration-filename`" from the previous step). After this function call, the `TDTState` variable is used to identify an open communication channel. In other words, the `TDTState` uniquely identifies one channel. If data is being transferred to or from more than one location, more `TDTState` variables need to be declared.

Read or Write Data

Calls to `tdt_read()` or `tdt_write()` perform the entire read or write operation for the specified variable. The user specifies the variable to be written (sent) and the identifier (`decl-name` from the name of element decl, see Sect. 2.5.1) which appears in the data description file.

```
tdt_read (ts, variable-name, "decl-name");
tdt_write (ts, variable-name, "decl-name");
```

All read operations are blocking. That is, the execution of the program will not proceed past a `tdt_read()` call if there is no data yet available.

Close the Channel

This step closes the named connection, frees up the socket for later use, and frees memory previously allocated for data structures associated with the connection.

```
tdt_close (ts).
```

2.6 Additional Technical Details

The TDT can be used in a number of programming languages. It has two core implementations in C and Python with interface functions supplied for Fortran and Java.

The library has been tested on a number of operating systems, namely Linux (the main development platform), AIX, *BSD, Mac OS X, Windows 98, NT, 2000 and XP and the Cygwin Unix environment for Windows.

Because the TDT has network sockets as one of it's underlying transfer protocols, communication over secure shell (SSH) connections is possible. Thus, for example, transparent communication between remote machines (heterogeneous coupling) is possible.

2.7 Conclusions and Perspectives

The TDT library is a small, self-contained library of functions for the transfer of data between programs written in diverse programming languages and across diverse operating systems. Its strengths lie in three main areas: firstly, its simple interface using read and write calls is intuitive to apply and requires minimal modifications to existing codes. Secondly, the ease of use of the TDT lends itself to rapid prototyping of coupled models either to try out a particular configuration or to develop an production-ready model. Thirdly, the flexibility of the TDT means that changes to the models being coupled are easily dealt with. For example, different communication channels such as network sockets, intermediate files or shared memory require no changes to existing model code and this means that trade-offs between the distribution of modules (using sockets) and performance (using shared memory) can be explored without major refactoring of software.

Although lacking advanced features such as interpolation, time-stepping and data marshalling, the TDT can be used as a basis for more advanced model coupling techniques. The ease with which the TDT can be employed makes it ideal for rapid prototyping of coupled solutions.

The TDT's user community is made up of several small groups from a variety of backgrounds. These include climate modelling at PIK, geophysical modelling, multi-agent social-biophysical modelling, both in production and for rapid-prototyping of coupled solutions. Also, the Dutch National Institute for Public Health and the Environment (RIVM) has developed the M modelling language for defining and visualising mathematical models. In conjunction with PIK, modifications were made to the M driver software to support communication of these models via TDT. Finally, the Bespoke Framework Generator (BFG) described in Chap. 7 supports the generation of wrappers for models which use TDT as their communication layer.

As an Open Source library of 6,000 lines of source code, released under the GPL it can be easily modified to include datatypes or transfer protocols not originally envisaged by the designers. It is also intended to be applicable to a user-generated suite of model coupling components which will eventually provide a more complex model coupling solution. At present there are no plans to add major new features to the TDT library, though active maintenance is on-going in response to bug reports and user requests for enhancements such as new data types.

Chapter 3
The Model Coupling Toolkit

Robert Jacob and Jay Larson

3.1 Introduction

The Model Coupling Toolkit (MCT) provides datatypes and methods for creating parallel couplers and parallel coupled models out of one or more models of physical systems. MCT handles common coupling tasks in a distributed memory parallel application.

Development of MCT began in 2000 under the U.S. Department of Energy's Office of Science and an 18-month program called the "Accelerated Climate Prediction Initiative (ACPI)—*Avante Garde*". A goal of this project was to enable the Community Climate System Model (CCSM, see Volume 1 of this series) to better exploit the then new microprocessor-based parallel computers. In particular, the coupler within CCSM (then called CSM) was not parallel while other components were developing limited parallelism. Communicating data required gathering and scattering all data to/from one processor creating a bottleneck to performance. The CCSM community had already settled on a hub-and-spoke design for future versions. CCSM was also using multiple numerical grids and contemplating more in the future. Thus MCT development focused on creating parallel data types, domain decomposition descriptors and parallel communication and interpolation methods for generic grids. Another goal was to provide a general solution to those problems, one that could be used in coupled applications from other fields. Climate models, and other coupled models, are supercomputing applications that require high performance to achieve simulation speed sufficient for their intended application. Supercomputers

R. Jacob (✉)
Argonne National Laboratory, Argonne USA
e-mail: jacob@mnsc.anl.gov

J. Larson
Research School of Computer Science, The Australian National University,
Canberra Australia
e-mail: larson@mnsc.anl.gov

S. Valcke et al., *Earth System Modelling – Volume 3*,
SpringerBriefs in Earth System Sciences, DOI: 10.1007/978-3-642-23360-9_3,
© The Author(s) 2012

come in many varieties so the codes must also be portable. Thus MCT was designed with an eye toward high-performance portability. Other software products developed for similar problems were surveyed but none could satisfy all these requirements.

MCT's designers had previous experience in developing a parallel coupler for a climate model (Jacob et al. 2001), which used a field regridding technique similar to the *exchange grid* (Sect. 5.4), and parallel analysis tools for data assimilation (Larson et al. 1998). The knowledge and techniques learned in those efforts greatly influenced the design of MCT. For example, the work with the Physical-space Statistical Analysis System (PSAS) encouraged the designers to write MCT entirely in Fortran90. Fortran90 has just enough object-oriented features to make it possible to write a class library and, because MCT's primary target applications are almost universally written in Fortran90, it was possible to avoid language interoperability issues. The experience with FOAM's coupler and examination of CCSM's coupler made it clear that MCT should focus on the hardest and most general problem: storage and communication of coupling fields in parallel between generic grids and decompositions.

MCT's first version was complete by November of 2002 and it was used as the basis for the coupler in CCSM3, released in June of 2004. This chapter presents an overview of the MCT programming philosophy and some of the features available in the toolkit.

3.2 Architectural Overview and Programming Philosophy

Elements of MCT's programming philosophy are found in many software applications. In general, MCT values flexibility over completeness and follows the "less is more" principle. For example, it was decided early on that MCT would not include a ready-to-use coupler application. From coupled models that existed at the time, it was clear that there were many choices regarding timestep, model sequencing, interpolation and physical flux calculations implemented in each model and the reasons behind these choices were more scientific then structural. In other words, the coupler is *part of the science* described in the model and must also be designed for readability: its inner workings must be easily comprehensible and it must also be reconfigurable as experimentation needs require. Trying to provide both a ready-to-use, universal coupler as well as readability and configurability are conflicting goals. The better way to simplify coupled model development would be to develop a set of programming guides that, with the help of a small but powerful set of software tools, would greatly reduce the development time of a new coupled model's coupler.

MCT is the small set of software tools. Architecturally, MCT is a Fortran90 library and calls to MCT are added to a user's program through F90 *module use* and compiled by linking with the library. MCT has three main pieces. At the lowest level is the Message Passing Environment Utilities (MPEU) which was part of PSAS. MPEU provides useful fortran-based utilities, e.g. sorting, multiprocessor stdout and stderr, and provides elements of MCT low-level classes. (MPEU can be used separately from

Fig. 3.1 The MCT
architecture

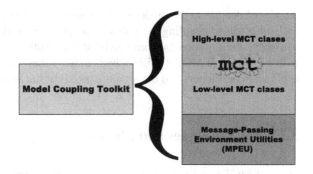

MCT.) MCT's low level classes handle the basics of coupling: a component model
registry, domain decomposition descriptors, communication schedulers for parallel
data transfer, a flexible and indexible data storage type, and parallel interpolation
algorithms (which use offline-generated weights). The higher level classes provide
useful functions within a coupler such as spatial averaging, time averaging, and
merging (Fig. 3.1).

A programming guide for coupled modeling would include the following elements:

- A model ideally should not assume it has sole control of the standard input and
  use named logical units instead.
- Similarly, a model may not be the only one outputting messages to standard out
  and text output should include a unique string identifying what model they came
  from or write to a named file.
- For parallel models that use the Message Passing Interface (MPI), a uniquely named
  communicator should be used instead of the general MPI_COMM_WORLD one.

At a higher level, models are easy to configure with several coupling strategies if
their design follows a pattern that comes from object oriented programming: i.e. each
model is separated into an initialization method and a run method (and possibly a
"finalize" method.) This same design pattern is advocated by other coupling software
(see e.g. ESMF in Sect. 6.5.1). The initialization method of a scientific application
often has many duties besides allocating memory; it must read in initial and boundary
value data and set values for important physical constants. With these standards,
the stand-alone model could have its topmost level rewritten as a relatively small
piece of software that calls the initialize, run and finalize methods of the model.
Once initialization methods are separated, they can be arranged to be each called in
sequence in the new coupled application before the run methods are called.

But for a legacy application, the effort to make an application callable from another
application can be as much or even more effort then building the complete coupled
model. For example, it can involve taking data structures which exist far down in
the calling tree and bringing them up to the run method if they contain coupling
state information. For the cases where it is not possible to make a legacy application callable from a new model or follow other of the guidelines above, MCT's
methods also support a programming model where MCT communication routines

are placed deep in to the model's calling sequences so they can communicate with each other (direct coupling) or with a central, user-created, coupler. Hybrids of these two programming models (some embedded and others separate) are also possible with MCT. Part of MCT's philosophy is to leave these important decisions up to the application developers and the scientific needs of the application.

3.3 MCT Datatypes and Methods

Some of MCT's methods are patterned after the Message Passing Interface (MPI), version 1. MCT uses MPI1 two-sided communication to move data between two models. It is expected, but not required, that the models in the coupled system are parallelized with MPI. A small, serial MPI replacement library is included in MCT so that MCT-coupled applications can be compiled without a full MPI library. Many MPI concepts such as *root* and *rank* are used in MCT with the same meaning. MCT is compatible with MPI2 but does not yet employ its one-sided communication model.

Because of the focus on parallelism one of the low-level classes in MCT is MCT_WORLD. MCT_WORLD contains an internal table of how many models are present in the coupled system and the local and global ranks of the processes they are running on. The result is a registry of all models running in the coupled application. This is initialized once at startup of the model and is referenced internally by MCT's communication routines.

MCT's communication routines provide a solution to what is sometimes called the "M × N problem" (Bertrand et al. 2006) which is the problem of how to transfer a distributed data object from a module running on M processes to another running on N processes. In a parallel coupled model, this would be the problem of getting values held in a one model's internal distributed data type to the other model's distributed type. This is the fundamental problem to solve when building a high-performance parallel coupled model. One of the core classes used to solve this problem is the GLOBALSEGMAP (GSMap) or *Global Segment Map*. The GSMAP is the MCT datatype for describing a decomposition of a numerical grid or the portion of the grid used in coupling. It provides a linearization of the grid. Each point must be assigned a unique integer value by the user. It is possible to describe any decomposition and grid, including unstructured, this way. Two simple decompositions of a grid are illustrated in Figs. 3.2 and 3.3. The choice of how exactly to number the grid points is left to the user.

The heart of the GSMAP datatype contains three arrays that are sized according to the total number of segments in the decomposed grid. These three arrays contain, for each index, the starting grid point number, the run length of the segment and the rank of the processor the segment resides on (the processor location). For the decomposition shown in Fig. 3.2, the GSMAP arrays would be *start*(1, 11):*length*(10, 10):*pe_loc*(0, 1) while for the decomposition in Fig. 3.3 the arrays would be *start*(1, 6, 3, 8, 11, 16, 13, 18):*length*(2, 2, 3, 3, 2, 2, 3, 3): *pe_loc*(0, 0, 1, 1, 2, 2, 3, 3). Each application needs to construct its own GSMAP. A copy of that GSMAP is available on

Fig. 3.2 A 2 × 1 decomposition of grid with 20 points. The points have been numbered

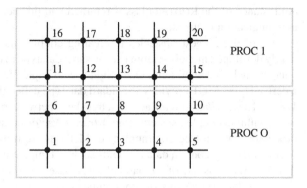

Fig. 3.3 A 2 × 1 decomposition of the same grid in Fig. 3.2

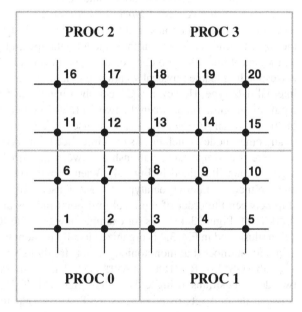

each processor after initialization. This does not present a memory scaling problem because the size of the three integer arrays grows only by the number of segments, not the number of points.

Once the GSMAP is initialized, MCT provides a method for determining and storing a communication schedule that describes for other routines how to transfer data between two different decompositions of a grid. The decompositions could be within the same set of processors, two completely disjoint sets of processors, or on overlapping sets of processors. This data type is called the ROUTER. The only input needed is either two GSMAPs, when a model has two decompositions of the same grid, or one GSMAP and the ID from MCTWORLD of the other component in the communication. The initialization of a router is like the old "handshake" performed by modems: the two models are learning how to exchange data with each other.

An instance of the ROUTER class is returned and this becomes an argument for the communication routines.

One of the biggest obstacles to creating coupled models is that two independently developed models seldom use the same datatype to describe the same physical concept such as a wind field. One might use a simple 3D array while the other uses a derived type. MCT provides a standard datatype called the ATTRVECT or *Attribute Vector*. The *Attribute Vector* acts as a translator between possibly very different data types within a coupled system. One *Attribute Vector* holds all of the data on a local processor that must be communicated. Methods are provided to access individual vectors that may contain data for fields such as temperature, wind or humidity. (The *Attribute Vector* is one of the parts of MCT based on MPEU).

All the data that needs to be communicated to another model must be copied in to an *Attribute Vector*. A second *Attribute Vector* on the other side of the communication receives the data and copies it in to that model's data type. The user must provide the logic to translate between their model datatype and the *Attribute Vector*. That is, they must know how to copy data from a location in the *Attribute Vector* to the corresponding (same physical location, e.g. latitude and longitude) location in their internal data type. However MCT can then handle all the communication for any parallel configuration between the two ATTRVECTS. Once a model can copy its data in and out of an *Attribute Vector*, it can use MCT routines to send data, in parallel, to any other model which has its own interface to the ATTRVECT.

There is an important relationship between the *Attribute Vector* and the global segment map illustrated in Fig. 3.4. When copying between the local ATTRVECT and the model's internal datatype, the MCT user must be mindful of the relationship between the order of the local grid point indices and the order of data in the ATTRVECT. Figure 3.4 shows for example how the "PROC 1" data of the decomposition illustrated in Fig. 3.2 is organized locally in monotonic increasing order of the grid point number (but monotonicity is not strictly necessary).

With a GSMAP, ROUTER and ATTRVECT defined, users can transfer data between two decompositions using either MCT_SEND/MCT_RECV, for transferring data between disjoint sets of processors, or REARRANGER for two decompositions on the same set of processors. The REARRANGER *init* method creates a ROUTER but keeps track of the points that are on the same physical memory and can be directly copied. The run method will then use message passing for those points which are on other processors and local copies for local points. REARRANGER is a collective operation while MCT_SEND/MCT_RECV follow the two-sided message passing model of MPI. Non-blocking versions of the later are also available.

Another common problem in creating coupled models is interpolating data from one grid representation to another. MCT currently supports interpolation as a linear transformation implementable as a sparse matrix-vector multiply. MCT provides the SPARSEMATRIX data type to hold matrix weights and the MATATTRVECMULT to perform the interpolation between data contained in an *Attribute Vector*. MCT does not currently provide methods for calculating the weights because there are potentially many ways to perform that crucial calculation. It is another science aspect that is best left to the coupled model developers. In practice, MCT typically

Fig. 3.4 The relationship between the local memory in an ATTRVECT and the grid point numbering contained in the GLOBALSEGMAP for this model. In this example, the ATTRVECT is being used to store three variables: temperature (T), east-west velocity (U) and specific humidity (Q)

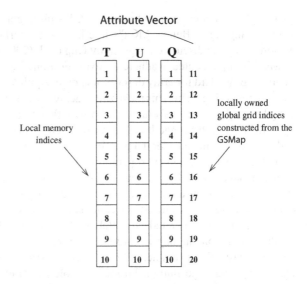

uses weights calculated by the Spherical Coordinate Remapping and Interpolation Package (SCRIP, Jones 1999).

The high-level classes in MCT include the ACCUMULATOR for time-averaging of data involved in coupling. This level also includes a set of transformation routines, including: time integration that computes ACCUMULATOR elements from ATTRVECT input; spatial integration routines for performing global sums of data in an ATTRVECT, which are often necessary for checking the results of interpolation; and support for weighting outputs from two or more models that must be sent to a third model as input—a process called *merging*. All of the classes mentioned here and above contain additional routines for functions such as query, and MCT datatype-specific support for standard parallel operations such as broadcast, gather and scatter. Further details can be found in Larson et al. (2005) and Jacob et al. (2005) and the MCT web site.[1]

3.4 MCT Multi-Language Interface

MCT's code base and programming model were based on Fortran modules and derived types, and this approach has been highly successful for Fortran-based scientific software. MCT's long-term goal is to create universally applicable code to support multiphysics and multiscale model construction for all areas of computational science and engineering—a goal that requires a strategy for language interoperability. MCT's philosophy of creating light coupling infrastructure dictated that language interoperability be separated from MCT's core functionality. A layer of interlan-

---

[1] http://www.mcs.anl.gov/mct.

guage "glue" was created that allows MCT's programming model to be exported to other languages. Below, MCT's interlanguage interfaces are briefly described. A more complete discussion is given by Ong et al. (2008).

MCT is implemented in adherence to the Fortran90 standard, which offers neither a complete standard for array descriptors nor an Application Programming Interface (API) for querying them. This makes interfacing codes in Fortran90 to other programming languages notoriously difficult. Fortran2003 offers a BINDC feature, but at the time MCT's interlanguage bindings were created, this feature was not sufficiently widely available to consider it a portable solution. The interfaces were implemented using the Babel language interoperability tool. Babel supports a variety of programming languages, including Fortran, C, C++, Python, and Java. The MCT multilingual interfaces were semi-automatically generated using Babel, with some intervention to implement references to MCT's code base. MCT APIs for C++ and Python were then automatically generated using Babel. The result is an MCT programming model for the destination languages that looks and feels natural to native programmers of these languages. For example, in MCT's Python API, Fortran module use is replaced with Python package import, Fortran variable declaration is replaced with a null constructor, and method invocation remains unchanged. The performance of MCT's C++ interfaces has been found to be close to their Fortran counterparts. The Python API is significantly slower, but still performs well enough to support a test implementation of the CCSM coupler in Python.

3.5 Conclusions and Perspectives

The most prominent user of MCT continues to be the Community Climate System Model and its developers. CCSM is one of the main climate models in the U.S. and is used for national assessment simulations of the effects of climate change. MCT is the basis of the "cpl6" coupler (Craig et al. 2005) that was part of the third version of the Community Climate System Model (Collins et al. 2006). It is also the basis of "cpl7" in the fourth version of CCSM and the first version of the Community Earth System Model[2] released in April and June of 2010, respectively. "cpl7" is much more flexible in that it allows several different ways for the components to interact (see Volume 1 of this series). The coupled model developers also have exposed more of MCT to its users, recognizing that MCT's simple concepts and functions help to clarify the flow of information in the coupler. Other researches at the National Center for Atmospheric Research are using MCT for more advanced coupled systems such as nesting a regional ocean model in a global ocean model.

Other users of MCT are notable because many were able to use MCT successfully and create new coupled applications while referring only to MCT's API document and some example programs distributed with the code. (MCT currently lacks a *Users' Guide*). These include a model for simulating hurricanes consisting of a weather

---

[2] http://www.cesm.ucar.edu

model coupled to a regional ocean model developed by researchers at NCAR and the NOAA Pacific Marine Environmental Laboratory and a system for studying coupled atmosphere-ocean processes along coastlines created by researchers at Oregon State University (Warner et al. 2008). Support for MCT users is provided by emailing the developers. The MCT development team has been 2–3 people throughout its history augmented by contributions to the code from users.

MCT's architecture has inspired the design of a coupling API (Larson & Norris, 2007) for the Common Component Architecture (CCA; Bernholdt et al. 2006). CCA's goal is to leverage component-based software engineering techniques to serve the computational science community. The coupling API comprises a set of composable components that can be assembled to create coupler applications, with components for the major types of data processing operations encountered in code coupling, including: component model registration; field, mesh, and domain decomposition data description; parallel data transfer and redistribution; linear transformations of field data; time averaging and integration of field and flux data; and merging. An MCT-based reference implementation of these coupling components is currently under development.

MCT continues to be improved. MCT has recently implemented a hybrid programming model (especially important for machines with low memory per node) using OpenMP in some of its routines. MCT has augmented its key communication routines so that they are scalable to tens of thousands of processors. In the future, MCT will also support internally some widely used methods for calculating interpolation weights and provide better support for unstructured grids. MCT continues to be supported by the US Department of Energy's Office of Science under its Scientific Discovery through Advanced Computing program. The MCT approach has been shown to be both flexible and easy to learn and should continue to grow in use.

References

Bernholdt DE, Allan BA, Armstrong R, Bertrand F, nneth Chiu K, Dahlgren TL, Damevski K, Elwasif WR, Epperly TGWE, Govindaraju M, Katz DS, Kohl JA, Krishnan, M, mfert GK, Larson JW, Lefantzi S, Lewis MJ, Malony AD, McInnes LC, Nieplocha J, Norris B, Parker SG, Ray J, ende SS, Windus TL, Zhou S (2006) A component architecture for high-performance scientific computing. Int J High Perform Comput Appl 20(2):163–202. doi 10.1177/1094342006064488

Bertrand F, Bramley R, Damevski K, Kohl J, Bernholdt D, Larson J, Sussman A (2006) Data redistribution and remote method invocation for coupled components. J Parallel Distributed Comput 66(7):931–946

Collins WD, Bitz CM, Blackmon ML, Bonan GB, Carton JA, Chang P, Doney SC, Hack JJ, Henderson TB, Kiehl JT, Large WG, McKenna DS, Santer BD, Smith RD (2006) The community climate system model: CCSM3. J Clim 19(11):2122–2143

Craig AP, Jacob RL, Kauffman BG, Bettge T, Larson J, Ong E, Ding, C, He H (2005) Cpl6: The new extensible high-performance parallel coupler for the community climate system model. Int J High Perform Comput Appl 19(3):309–327

Jacob R, Schafer C, Foster I, Tobis M, Anderson J (2001) Computational design and performance of the fast ocean atmosphere model. In: Alexandrov VN, Dongarra JJ, Tan CJK (eds).In: Proceedings of 2001 international conference on computational science, Springer-Verlag, pp 175–184

Jacob R, Larson J, Ong E (2005) MxN communication and parallel interpolation in CCSM3 using the model coupling tookit. Int J High Perform Comput Appl 19(3):293–307

Jones P (1999) First- and second-order conservative remapping schemes for grids in spherical coordinates. Monthly Weather Rev 127:2204–2210

Larson J, Norris B (2007) Component specification for parallel coupling infrastructure. In: Gervasi O, Gavrilova M (eds) Proceedings of the international conference on computational science and its applications (ICCSA 2007) Lecture notes in computer science, vol 4707. Springer-Verlag, pp 55–68

Larson J, Jacob R, Ong E (2005) The model coupling toolkit: a new Fortran90 toolkit for building multi-physics parallel coupled models. Int J High Perform Comput Appl 19(3):277–292

Larson JW, Guo J, Gaspari G, da Silva A, Lyster PM (1998) Documentation of the physical-space statistical analysis system (PSAS) Part III: The software implementation, Technical report DAO Office Note 98-05, NASA/Goddard Space Flight Center, Greenbelt, Maryland

Ong ET, Larson JW, Norris B, Jacob RL, Tobis M, Steder M (2008) A multilingual programming model for coupled systems. Int J Multiscale Comput Eng 6(1):39–51

Warner J, Perlin N, Skyllingstad E (2008) Using the model coupling toolkit to couple earth system models. Environ Modelling Softw 23(10–11):1240–1249

Chapter 4
The OASIS Coupler

Sophie Valcke and René Redler

4.1 Introduction

In 1991, CERFACS was commissioned by the French climate modelling commu-
nity to perform the technical assembling of an ocean General Circulation Model
(GCM), Océan Parallélisé (OPA) developed by the Laboratoire d'Océanographie
Dynamique et de Climatologie (LODYC), and two different atmospheric GCMs,
Action de Recherche Petite Echelle Grande Echelle (ARPEGE) and the Laboratoire de Météorologie Dynamique zoom (LMDz) model developed respectively by
Météo-France and the Laboratoire de Météorologie Dynamique (LMD). Two years
later, a first version of the OASIS coupler was distributed to the community and
used in a 10-year coupled integration of the tropical Pacific (Terray et al. 1995). At
the time, the communication, i.e. the exchange of coupling fields, was ensured with
CRAY named pipes and ASCII files. Between 1995 and 2000, alternative communication techniques based on the Parallel Virtual Machine (PVM, http://www.csm.
ornl.gov/pvm/), UNIX System V Interprocess Communication (SVIPC), NEC SX
global memory communication and on the Message Passing Interface (MPI) were
introduced while the community of users was steadily and rapidly growing in Europe
but also in Australia and in the USA.

From 2001 until 2004, the development of OASIS benefited from an important
support from the European Commission in the framework of the PRISM project[1]
(Valcke et al. 2006, see also Volume 1 of this series). The collaboration with NEC

---

[1] http://prism.enes.org

S. Valcke (✉)
CERFACS, Av. Coriolis 42, 31057 Toulouse Cedex 01, France
e-mail: sophie.valcke@cerfacs.fr

R. Redler
Max-Planck-Institut für Meteorologie, Bundesstraße 53, 20146 Hamburg, Germany
e-mail: rene.redler@zmaw.de

S. Valcke et al., *Earth System Modelling – Volume 3*,
SpringerBriefs in Earth System Sciences, DOI: 10.1007/978-3-642-23360-9_4,
© The Author(s) 2012

Laboratories Europe-IT Research Division (NLE-IT) and the French Centre National de la Recherche Scientifique (CNRS) originated during that period.

Today, both the widely used OASIS3 coupler (Valcke 2006) and the new fully parallel OASIS4 coupler (Redler et al. 2010) are available and used by about 35 different climate modelling groups around the world. In the next paragraphs, technical details about the coupler functionality are provided, emphasizing the similarities and differences between OASIS3 and OASIS4.

4.2 Architectural Overview

After an initial period of investigation, it was initially decided that the technical coupling layer between the ocean and atmosphere components should take the form of an external coupler, i.e. a separate executable performing the regridding of the coupling fields, and a coupling library linked to the components performing the coupling exchanges. The components remain separate executables with their main characteristics practically unchanged with respect to the uncoupled mode. The coupling options for a particular run (e.g. the interpolations to apply on the coupling fields) are defined by the user in an external configuration file. These choices ensure that only minimal adaptation has to be done in the existing codes to couple them through OASIS and also that the coupled set-up can easily be changed without modifying the code of the components themselves. Low intrusiveness, modularity and portability are therefore OASIS main design criteria.

During the first years, as the coupling was at the time involving only a relatively small number of 2D coupling fields at the air-sea interface, efficiency was not considered a major criteria. The OASIS3 version is the direct evolution of the developments done in CERFACS since 1991 following these design principles. However, as the climate modelling community was progressively targeting higher resolution climate simulations run on massively parallel platforms with coupling exchanges involving a higher number of (possibly 3D) coupling fields at a higher coupling frequency, the development of a new fully parallel coupler, OASIS4, started during the PRISM project. Parallelism and efficiency drove OASIS4 developments, at the same time keeping in its design the concepts of portability and flexibility that made the success of OASIS3.

Both OASIS3 and OASIS4 are portable set of Fortran and C routines. After compilation, they form a separate executable performing driving and regridding tasks and a model coupling interface library, the PSMILe, that needs to be linked to and used by the component models.

4.3 Coupling Configuration

At run time, the OASIS Driver first reads the coupled run configuration defined by the user before the run and distributes the corresponding information to the different component model PSMILes. This user-defined configuration contains all coupling options for a particular coupled run, e.g. the duration of the run, the component models, and for each coupling exchange a symbolic description of the source and target, the exchange period, regridding and other transformations. During the run, the Driver-Transformer executable and the component model PSMILes perform appropriate exchanges based on this configuration.

With OASIS3, the configuration information is contained in a text file following a specific format while with OASIS4 it is provided in Extensible Markup Language (XML, http://www.w3.org/XML/) files; a Graphical User Interface (GUI) is currently being developed to facilitate the creation of those XML files.

4.4 Process Management

In a coupled run using OASIS3 or OASIS4, the component models generally remain separate executables with main characteristics, such the general code structure or the memory management, untouched with respect to the uncoupled mode. OASIS supports two ways of starting the executables of the coupled application. If a complete implementation of the MPI2 standard (Gropp et al. 1998) is available, only the OASIS Driver has to be started by the user. All remaining component executables are then launched by the OASIS Driver at the beginning of the run using the MPI2 MPI_Comm_Spawn functionality. If only MPI1 (Snir et al. 1998) is available, the OASIS Driver and the component model executables must be all started at once in the job script in an "multiple program multiple data" (MPMD) mode. The advantage of the MPI2 approach is that each component keeps its own internal communication context unchanged with respect to the standalone mode, whereas in the MPI1 approach, OASIS needs to recreate a component model communicator that must be used by the component model for its own internal parallelisation. In both cases, all component models are necessarily integrated from the beginning to the end of the run, and each coupling field is exchanged at a fixed frequency defined in the configuration file for the whole run.

4.5 Communication: The OASIS PSMILe Library

To communicate with other component models or to perform I/O, a component model needs to call few specific OASIS PSMILe routines. The PSMILe API function calls for both OASIS3 and OASIS4 can be split into three phases. The first phase includes calls for the coupling initialisation, the definition of the grids (i.e. the grid point and corner longitude and latitude), the description of the local partition in a global index space, and the coupling field declaration; the second phase comprises receiving and sending of the coupling fields (by calling respectively a *prism_get* or a *prism_put* routine) usually implemented in the model time stepping loop, while the third phase terminates the coupling.

The OASIS4 PSMILe API was kept as close as possible to the OASIS3 PSMILe API; this should ensure a smooth and progressive transition between the OASIS3 and OASIS4. The main difference between the OASIS3 and OASIS4 PSMILe API remains in the grid definition; while with OASIS4 the description of the local partition of the grid covered by the local process is mandatory, with OASIS3 the global grid definition can either be provided by the component master process through the PSMILe API or by the user in a NetCDF (http://www.unidata.ucar.edu/software/netcdf/) file constructed before the run.

For both OASIS3 and OASIS4, the sending and receiving of data is managed by the PSMILe below the *prism_get* and *prism_put* calls, following a principle of "end-point" data exchange. When producing data, no assumption is made in the source component code concerning which other component will consume these data or whether they will be written to a file, and at which frequency; likewise, when asking for data, a target component does not know which other component model produces them or whether they are read in from a file. The target or the source (another component model or a file) for each field is defined by the user in the configuration file and the coupling exchanges and/or the I/O actions take place according to the user external specifications. This implies in particular that the switch between the coupled mode and the forced mode is totally transparent for the component model. MPI is used for coupling exchanges, while I/O actions are based on GFDL mpp_io library (Balaji 2001).

Furthermore, the *prism_get* and *prism_put* routines can be placed anywhere in the source and target code and possibly at different locations for the different coupling fields. These routines can be called by the model at each timestep. The actual date at which the call is valid is given as argument and the sending/receiving is actually performed only if the date corresponds to a time at which it should be activated, given the field coupling or I/O frequency indicated by the user in the configuration file; a change in the coupling or I/O frequency is therefore also totally transparent for the component model itself.

Both OASIS3 and OASIS4 PSMILe support parallel communication in the sense that each process of a parallel model can send or receive its local part of the field. With the OASIS3 PSMILe, the different local parts of the field are sent to the OASIS3 Transformer which gathers the whole coupling field, transforms or regrids it, and redistributes it to the target component model processes. With the last release of OASIS3, it is now possible to run the Transformer on more than one process, each process treating a subset of the complete coupling fields; the result is a parallelisation of OASIS3 on a field-per-field basis. With the OASIS4 PSMILe, the communication is more efficient as the communication pattern between the source and target processes is based on the intersection of the local domains covered by each source and target component process; therefore, only the useful part of the coupling field is extracted and transferred, either directly between the models (when only repartitioning is needed) or via the parallel Transformer (when repartitioning and regridding are needed).

4.6 Coupling Field Transformation and Regridding

For each coupling exchange, the OASIS Transformer receives the source coupling field (or part of it) from the source model, performs the transformations and regridding needed to express the source field on the grid of the target model, and sends the transformed field (or part of it) to the target model.

4.6.1 Transformation and Regridding in OASIS3

As stated above, with OASIS3, the transformation is done by the Transformer mono-process executable on the whole coupling field after its gathering in the Transformer memory. The neighbourhood search, i.e. the determination for each target point of the source points that contribute to the calculation of its regridded value, and the corresponding weight calculation is done by the Transformer at the beginning of the run considering the whole source grid. Different transformations on 2D coupling fields in the Earth spherical coordinate system are available for grids that are regular in longitude and latitude, stretched, rotated, Gaussian reduced, and unstructured:

- time accumulation or averaging
- "correction" with external data read from a file
- linear combination with other coupling fields
- addition or multiplication by a scalar
- nearest-neighbour, Gaussian-weighted, bilinear, bicubic 2D interpolations
- 2D conservative remapping (i.e. the contribution of each source cell is proportional to the fraction of the target cell it intersects);
- any user-defined regridding (the weights and addresses are pre-defined by the user in an external file)
- global conservation
- creation of subgrid scale variability (when regridding from low to high resolution)

The interpolations and the conservative remapping are taken from the Spherical Coordinate Remapping and Interpolation Package (SCRIP) library (Jones 1999). OASIS3 also supports interpolation of vector fields with the projection of the two vector components in a Cartesian coordinate system, interpolation of the resulting three Cartesian components, and projection back in the spherical coordinate system. OASIS3 can also be used in the *interpolator-only* mode to transform and regrid fields contained in files without running any model.

4.6.2 *Transformation and Regridding in OASIS4*

During the run, the OASIS4 parallel Transformer manages the transformation and regridding of 2D or 3D coupling fields. The Transformer performs only the weight calculation and the regridding per se while the neighbourhood search is performed in parallel in the source PSMILe; this ensures that only the useful part of the coupling field is extracted and transferred.

In a simple implementation of a neighbourhood search algorithm, such as the one used in OASIS3, the "neighbours" of a target grid with M points can be identified by comparing their distance to all N source points. While such an approach of the order (M × N) is still justified for relatively small problem sizes, it may become very costly for high resolution grids. For OASIS4, a more efficient algorithm is implemented. In a first step, envelopes of the grid partitions residing on each process are defined, exchanged between source and target processes and intersections are identified. For each intersection, the list of target points is sent to the source process; a grid hierarchy, similar to what is used by a multigrid algorithm, is established on the source side with a refinement factor of two. In OASIS4, this hierarchy is used to identify the source cell containing the projection of each target point. From this step onwards, the identification of the neighbours involves only local operations in the grid point space. The great advantage of the "multigrid" algorithm is that it is of the order of (M × log N) and thus only weakly dependent on the source grid size; e.g. only one additional level is introduced in the source grid hierarchy if the source grid size is doubled.

When the source grid is partitioned, some neighbours of a target grid point which projection is contained into one source partition near its border can in fact be located on an adjacent partition. OASIS4 performs this additional search step, called the parallel "global" search, into which the source neighbours are also searched in adjacent partitions. If neighbour source points are found in adjacent partitions, the full information about those source points is returned to the process that has initiated the search. The global search therefore ensures that the regridding result is independent of the source partitioning.

At the end of the PSMILe neighbourhood search, each source process holds different lists, each list containing the information about the target points located in the intersection of a target process domain with its local domain and about the source neighbour points needed for the regridding of these target points. These lists are initially equally distributed over the Transformer processes, resulting in an effective parallelisation of the Transformer over the lists.

During the simulation time stepping, the OASIS4 parallel Transformer can be assimilated to an automate that reacts to what is demanded by the different component model PSMILes. During the exchange phase, each Transformer process receives from the source PSMILe the grid point field values (transferred from the source component with a *prism_put* call), calculates the regridding weights if it is the first exchange, and applies the weights. The data are sent upon request from the respective target process (i.e. when a *prism_get* is called in the target component code). The OASIS4

Transformer therefore acts as a parallel buffer in which the transformations take place.

In OASIS4, the following transformations are available for 2D and 3D coupling fields in the Earth spherical coordinate system for grids that are regular in longitude and latitude, stretched, rotated, or Gaussian reduced (unstructured grids are not supported yet):

- time accumulation or averaging
- addition or multiplication by a scalar
- gathering/scattering (required when the grid definition includes all masked and non masked points but when the coupling field itself gathers only non masked points)
- 2D nearest-neighbour, Gaussian-weighted, bilinear, bicubic interpolations
- 3D nearest-neighbour, Gaussian-weighted, trilinear interpolations
- 2D conservative remapping

The parallel global search is implemented for all grids supported and for all interpolations. As in OASIS3, the 2D algorithms are taken from the Spherical Coordinate Remapping and Interpolation Package (SCRIP) library (Jones 1999). The 3D algorithms are 3D extensions of the 2D SCRIP algorithms. OASIS4 also supports user-defined regridding but not vector interpolation.

4.7 Performances

First tests of scalability and performance of OASIS4 are detailed in Redler et al. (2010). For OASIS3, it is interesting to note that it has been used recently in few high-resolution coupled simulations without introducing significant overhead in the simulation elapse time even if it is parallelised only a field-per-field basis:

- At the UK Met Office, OASIS3 is used to couple the atmospheric Unified Model (UM) with a horizontal resolution of 432×325 grid points and 85 vertical levels to the ocean NEMO model including the CICE sea ice model at a horizontal resolution of $0.25°(1.5 \times 10^6$ grid points) and 75 depth levels. The coupling exchanges are performed every 3 h and the coupled model is run on an IBM power6 with 192 cpus for the UM, 88 cpus for NEMO and 8 cpus OASIS3. In this case, less than 2% overhead in the simulation elapse time was observed.
- OASIS3 is also used in the high-resolution version of the IPSL Earth System Model in Paris (France), coupling the LMDz atmospheric model with 589,000 points horizontally ($\frac{1}{3}°$) and 39 vertical levels to the NEMO ocean model in the ORCA0.25 configuration (1.5×10^6 points) and 75 depth levels on the CINES SGI ALTIX ICE. The coupling exchanges are performed every 2 h. The coupled system uses up to 2191 cpus with 2048 for LMDz, 120 for NEMO and 23 for OASIS3.

- Recently, the resolution of EC-Earth [2] coupled model was increased to T799 (about 25 kms, 843,000 points) and 62 vertical levels for the atmospheric model IFS and to ORCA0.25 configuration (1.5×10^6 points) and 45 depth levels for NEMO ocean model. This was run on the Ekman cluster (1268 nodes of 2 quadripro AMD Opteron, i.e. a total of 10,144 cores) with different numbers of cores for each component and OASIS3. Even if OASIS3 elapse time is non negligible when it runs in mono-processor mode (observed overhead was 13.4 and 11% when the components of the coupled model were run on the O(500) and O(1000) cores), it was also observed that when the parallelism of OASIS3 increases (going from 1 to 10 cores), OASIS3 elapse time decreases and its cost can almost be "hidden" in the component imbalance. In this case, the coupling overhead went down from 13.4 to 3% in a configuration with 512 cores for IFS and 128 cores for NEMO and from 11 to 1.3% in a configuration with 800 cores for IFS and 256 cores for NEMO.

In conclusion, OASIS3 parallelisation on a field-per-field basis can be an efficient way to reduce the coupling overhead. Of course, this way of hiding the cost of OASIS3 works only if there is some imbalance of the components elapse time which allows OASIS3 to interpolate the fields when the fastest component waits for the slowest. If the components were perfectly load balanced, then OASIS3 cost, even if lower when it is run in parallel, would be directly added to the component model elapse time.

4.8 User Community

Since the first version released in 1993 and used at CERFACS, Météo-France and IPSL in France, the number of OASIS users steadily increased and reaches today a community of about 35 climate modelling groups. OASIS success up to now can be explained by its great flexibility, the active support offered by the development team to the users, and the great care taken to constantly integrate the community developments in the official version.

OASIS3 in particular is used today by many different climate modelling groups in Europe, Australia, Asia and North America among which Météo-France and IPSL in France, the European Centre for Medium range Weather Forecasts (ECMWF), the Max-Planck Institute for Meteorology (MPI-M) in Germany, the Met Office and the National Centre for Atmospheric Science (NCAS) in the UK, the "Koninklijk Nederlands Meteorologisch Instituut" (KNMI) in the Netherlands, the Swedish Meteorological and Hydrological Institute (SMHI) in Sweden, the "Istituto Nazionale di Geofisica e Vulcanologia" (INGV) and the "Ente Nazionale per le Nuove tecnologie, l'Energia el Ambiente" (ENEA) in Italy, the Bureau of Meteorology (BoM) and the Commonwealth Scientific and Industrial Research Organisation (CSIRO) in

---

[2] http://ecearth.knmi.nl/

Australia, the Université du Québec à Montréal and Service Météorologique du Canada, the Jet Propulsion Laboratory (JPL).

The current user community of OASIS4 is of course much smaller but use of OASIS4 has already shown promising results in different configurations. A first version of OASIS4 was used to realize a coupling between the Modular Ocean Model version4 (MOM4) ocean model and a pseudo atmosphere model at GFDL in Princeton (USA), and with pseudo models to interpolate data onto high resolution grids at the Leibniz Institute of Marine Sciences at the University of Kiel (IFM-GEOMAR) in Germany. OASIS4 is also used for 3D coupling between atmosphere and atmospheric chemistry models at ECMWF, KNMI and Météo-France in the framework of the EU GEMS project (Hollingsworth et al. 2008) and EU MACC project. Currently, OASIS4 is used at SMHI and at BoM in Australia for regional ocean-atmosphere coupling, and at the Alfred Wegener Institute (AWI, Bremerhaven, Germany) for global ocean-atmosphere coupling. OASIS4 is also currently under evaluation at CERFACS and MPI-M for global ocean-atmosphere coupling.

4.9 Conclusions and Perspectives

As detailed above, the OASIS coupler is software allowing synchronized exchanges of coupling information between numerical codes representing different components of the climate system. At run-time, OASIS is a separate application that performs the interpolation of the coupling fields and a communication library linked to the component models. Today OASIS is used by about 35 different climate modelling groups in Europe, Australia, Asia and North America.

Two versions of the coupler are currently available. OASIS3 is the direct evolution of the 2D coupler developed since more than 15 years at CERFACS. OASIS4 is a newer fully parallel 3D coupler first developed in collaboration with NLE-IT, SGI and the French CNRS. OASIS3 is stable and well debugged but it is more limited than OASIS4, which however still needs some validation, especially in the fully parallel cases. The OASIS4 PSMILe API was kept as close as possible to OASIS3 PSMILe API; this should ensure a smooth and progressive transition between OASIS3 and OASIS4 use in the climate modelling community.

Within the framework of funded projects, work continues to extend OASIS existing functionality and to establish comprehensive services around OASIS through a portal offering documentation, user guides, tutorial, FAQs, user forum and tips for best practices. One example of such an initiative is the InfraStructure for the European Network for Earth System Modelling (IS-ENES) project, started in March 2009 for 4 years in which a fruitful collaboration with the Deutsches Klimarechenzentrum GmbH (DKRZ) is currently taking place. In addition to the tasks already mentioned above, personal user support is provided within the IS-ENES project to the climate modelling community to assemble new models coupled with the OASIS3 or OASIS4 coupler, or migrating from OASIS3 to the fully parallel OASIS4.

Capitalizing about 35 person-years of work and being used by about 35 modelling groups, OASIS is a great example of successful software sharing, as the 1.0 person-year/group it represents is certainly much less than the time it would have taken for each group to develop its own coupler.

References

Balaji V (2001) Parallel numerical kernels for climate models. In: Proceedings of ECMWF tera-Computing workshop 2001, World Scientific Press

Gropp W, Huss-Lederman S, Lumsdaine A, Lusk E, Nitzberg B, Saphir W, Snir M (1998) MPI—The complete reference, vol 2, The MPI Extensions, MIT Press, Cambridge

Hollingsworth A, Engelen R, Textor C, Benedetti A, Boucher O, Chevallier F, Dethof A, Elbern H., Eskes H, Flemming J, Granier C, Kaiser J, Morcrette J, Rayner P, Peuch, V, Rouil L, Schultz M, Simmons A, Consortium TG (2008) Toward a monitoring and forecasting system for atmospheric composition: the gems project. Bulletin of the American Meteorological Society 89:1147–1164

Jones P (1999) First- and second-order conservative remapping schemes for grids in spherical coordinates. Month Weath Review 127:2204–2210

Valcke S, Guilyardi E, Larsson C (2006) Prism and enes: a european approach to earth system modelling. Concurrency and Computatation: Practice and Experience 18(2):231–245

Valcke S (2006) OASIS3 User Guide (prism_2-5) (pp 60). Technical Report TR/CMGC/06/73 CERFACS: Toulouse, France

Snir M, Otto S, Huss-Lederman S, Walker D, Dongarra J (1998) MPI—The complete reference, vol 1, The MPI Core (2nd ed), MIT Press, Cambridge

Terray L, Thual O, Belamari S, Déqué M, Dandin P, Delecluse P, Levy C (1995) Climatology and interannual variability simulated by the arpege-opa coupled model. Clim Dyn 11:487–505

Redler R, Valcke S, Ritzdorf H (2010) Oasis4 - a coupling software for next generation earth system modelling. Geosci Model Develop 3:87–104. doi 10.5194/gmd-3-87-2010

Chapter 5
The Flexible Modeling System

V. Balaji

5.1 Introduction: The Emergence of Modeling Frameworks

In climate research, with the increased emphasis on detailed representation of individual physical processes governing the climate, the construction of a model has come to require large teams working in concert, with individual sub-groups each specializing in a different component of the climate system, such as the ocean circulation, the biosphere, land hydrology, radiative transfer and chemistry, and so on. The development of model code now requires teams to be able to contribute components to an overall coupled system, with no single kernel of researchers mastering the whole. This may be called the *distributed development model*, in contrast with the monolithic small-team model development process of earlier decades.

A simultaneous trend is the increase in hardware and software complexity in high-performance computing, as we shift toward the use of scalable computing architectures. Scalable architectures come in several varieties, including shared-memory parallel vector systems or distributed memory massively-parallel systems. The individual computing elements themselves can embody complex memory hierarchies, especially in the recent trend toward multi-core and many-core systems, co-processors and accelerators. To facilitate sharing of code and development costs across multiple institutions, it is necessary to abstract away the details of the underlying architecture and provide a uniform programming model across different scalable and uniprocessor architectures.

The GFDL Flexible Modeling System (FMS)[1] is an early example of a *modeling framework*, a comprehensive programming model and toolkit for the construction of coupled climate models. The climate system is composed of hierarchies of inter-

---

[1] http://www.gfdl.noaa.gov/fms

V. Balaji (✉)
Princeton University, New Jersey USA
e-mail: balaji@princeton.edu

S. Valcke et al., *Earth System Modelling – Volume 3*,
SpringerBriefs in Earth System Sciences, DOI: 10.1007/978-3-642-23360-9_5,

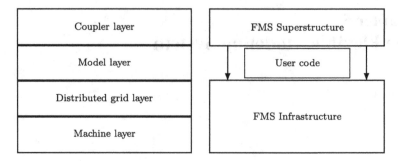

Fig. 5.1 The architecture of FMS

acting physical components, and it is natural to think of constructing models of such systems out of interacting *code* components. Component-based design of model codes is based on defining standards for what a component interface looks like: a community of scientists and developers that agree to adhere to a given *standard component interface (SCI)* set can then distribute development amongst themselves, confident that their own independently developed components will interact correctly with others within the same modeling framework. The FMS coupler is a *domain-specific* SCI: it is written quite narrowly to support ESMs.

In Sect. 5.2, we provide an architectural overview of FMS while in Sect. 5.3 we present the specific FMS architecture accommodating the physical requirements of the Earth system. In Sect. 5.4, we detail a central technology of the FMS coupler developed to meet these requirements, the *exchange grid*. Finally, in Sect. 5.5, we describe an additional key feature, i.e. the support for data assimilation. Section 5.6 concludes this chapter with a summary and a description of FMS users.

5.2 Architectural Overview

The schematic description of the FMS architecture is shown in Fig. 5.1. This "sandwich" architecture is fairly typical of such frameworks. User code, that is to say a set of routines expressing scientific algorithms, is written following the conventions of a standard infrastructure layer that provides useful and common technical services such as I/O, exception handling, and most importantly, operations on distributed grids and fields. Such standard high level expressions of parallelism, independent of the underlying hardware architecture, and uniformly expressed on all platforms, are an area of keen research interest. An overview of the field is given in volume 2 of this series. This chapter concerns itself instead with the *superstructure* layer of Fig. 5.1.

The FMS coupler is designed to address the question of how different components of the Earth system, say atmosphere and ocean, are discretized. Earlier generations of climate models used the *same* discretization, or simple integer refinement, for all components: thus, data exchange between components was a relatively simple point-to-point exchange. But any limitation on resolution of one component necessarily imposed itself on the other as well. Now it is increasingly common for each model component to make independent discretization choices appropriate to the particular physical component being modeled. In this case, how is, say a sea surface temperature from an ocean model made available to an atmosphere model that will use it as a boundary condition *on a different spatial grid*?

This is the *regridding problem*, subject to the following constraints when specialized to Earth system models:

- Quantities must capable of being *globally conserved*: if there is a flux of a quantity across an interface, it must be passed conservatively from one component to the other. This consideration is less stringent when modeling weather or short-term (intraseasonal to interannual) climate variability, but very important in models of secular climate change, where integration times can be $\mathcal{O}(10^6) - \mathcal{O}(10^8)$ timesteps.
- The numerics of the flux exchange must be stable, so that no limitation on the individual component timestep is imposed by the boundary flux computation itself.
- There must be no restrictions on the discretization of a component model. In particular, resolution or alignment of coordinate lines cannot be externally imposed. This also implies a requirement for *higher-order interpolation* schemes, as low-order schemes work poorly between grids with a highly skewed resolution ratio. Higher-order schemes may require that not only fluxes, but their higher-order spatial derivatives as well, be made available to regridding algorithms. The independent discretization requirement extends to the time axis: component models may have independent timesteps. (However, we do have a current restriction that a coupling timestep be an integral multiple of any individual model timestep, and thus, timesteps of exchanging components may not be co-prime).
- The exchange must take place in a manner consistent with all physical processes occurring near the component surface. This requirement is highlighted because of the unique physical processes invoked near the planetary surface: in the atmospheric and oceanic boundary layers, as well as in sea ice and the land surface, both biosphere and hydrology.
- Finally, we require computational efficiency on parallel hardware: a solution that is not rate-limiting at the scalability limits of individual model components. Components may be scheduled serially or concurrently between coupling events.

The specificity of the problem that the FMS coupler is designed to address distinguishes it from more general component frameworks such as the Earth System Modeling Framework (ESMF, see Chap. 6). Unlike an ESMF application, which can be recursively constituted out of components performing any function, the FMS coupler recognizes only a few components that may be on independent grids: an *atmosphere*, an *ocean surface*, a *land surface*, and an *ocean*. The ocean surface also represents the sea ice. Any other components inherit a grid from these, e.g. atmospheric physics and chemistry from the atmosphere; terrestrial biosphere, river and land ice components from the land surface; marine biogeochemistry from the ocean.

5.3 Physical Architecture of FMS Coupled System

As said above, the SCI for FMS is not phrased in terms of a generic "component" as in ESMF. Instead, there are interfaces or "slots" for each of the specific components listed above. For instance, an ocean model would encode its state in terms of specific data structures to hold the fields it exchanges with other components, called `ocean_boundary_type` and `ocean_data_type`. It must provide calls named `ocean_model_init` and `ocean_model_end` for initialization and termination, and a routine called `update_ocean_model` that steps the model forward for one coupling timestep. These calls all have a specific syntax. Each slot also includes the possibility of a null or "stub" component if that component is not needed, as well as a "data" component (where for instance the ocean is replaced by a dataset). In addition we provide a `data_override` capability for fine-tuned sensitivity studies, where individual fields in the model can be overridden by a dataset.

CodeBlock 1 shows an example of such a data structure, used to exchange data between ice and ocean components. The type is composed principally of a number of 2D surface fields, and the variable `xtype` which encodes the type of exchange.

```
type, public :: ice_ocean_boundary_type
real :: u_flux(:,:) ! wind stress (Pa)
real :: v_flux(:,:) ! wind stress (Pa)
real :: t_flux(:,:) ! sensible heat flux (W/m2)
real :: q_flux(:,:) ! specific humidity flux (kg/m2/s)
real :: salt_flux(:,:) ! salt flux (kg/m2/s)
real :: lw_flux(:,:) ! long wave radiation (W/m2)
real :: sw_flux_vis_dir(:,:) ! direct visible sw radiation (W/m2)
real :: sw_flux_vis_dif(:,:) ! diffuse visible sw radiation (W/m2)
real :: sw_flux_nir_dir(:,:) ! direct near IR sw radiation (W/m2)
real :: sw_flux_nir_dif(:,:) ! diffuse near IR sw radiation (W/m2)
real :: lprec(:,:) ! mass flux of liquid precip (kg/m2/s)
real :: fprec(:,:) ! mass flux of frozen precip (kg/m2/s)
real :: runoff(:,:) ! mass flux of liquid runoff (kg/m2/s)
real :: calving(:,:) ! mass flux of frozen runoff (kg/m2/s)
real :: runoff_hflx(:,:) ! heat flux of liquid land water (W/m2)
real :: calving_hflx(:,:) ! heat flux of frozen land water (W/m2)
real :: p(:,:) ! pressure of sea ice and atmosphere (Pa)
integer :: xtype ! REGRID, REDIST or DIRECT
type(coupler_2d_bc_type) :: fluxes ! additional tracers
end type ice_ocean_boundary_type
```

Note that the names of the exchange fields are hardcoded into the type. A scientist wishing to add a new exchange field does so here, and modifies the exchange code accordingly. We could have of course chosen to encode a list of anonymous exchange fields, naming the actual physical fields through external metadata. This makes the code more abstract, but easier to change the physical content of an exchange. We are always faced with such choices in code design. In the case of FMS and its user community, we found that the contents are not changed often, and hardcoding the physical contents to be the appropriate choice.

The exchange type variable xtype takes one of the values REGRID, REDIST, or DIRECT, which instructs the exchange software as to how to perform the exchange. REGRID is the most general: ice and ocean have completely independent grids. REDIST implies identical *physical* grids, but different parallel decomposition: data must be moved among processors but no regridding is needed. DIRECT implies a simple copy: both components have identical grids and decompositions.

Fluxes at the surface often need to be treated using an implicit timestep. Vertical diffusion in an atmospheric model is generally treated implicitly, and stability is enhanced by computing the flux at the surface implicitly along with the diffusive fluxes in the interior. Simultaneously we must allow for the possibility that the surface layers in the land or sea ice have vanishingly small heat capacity. This feature is key in the design of the FMS coupler. Consider simple vertical diffusion of temperature in a coupled atmosphere-land system:

$$\frac{\partial T}{\partial t} = -K \frac{\partial^2 T}{\partial z^2} \tag{5.1}$$

$$\Rightarrow \frac{T_k^{n+1} - T_k^n}{\delta t} = -K \frac{T_{k+1}^{n+1} + T_{k-1}^{n+1} - 2T_k^{n+1}}{\delta z^2} \tag{5.2}$$

$$\Rightarrow \mathbf{A T}^{n+1} = \mathbf{T}^n \tag{5.3}$$

This is a tridiagonal matrix inversion which can be solved relatively efficiently using an up-down sweep, as shown in Fig. 5.2. The problem is that some of the layers are in the atmosphere and others are in the land. Moreover, if the components are on independent grids, the key flux computation at the surface, to which the whole calculation is exquisitely sensitive, is a physical process (e.g Monin & Obukhov, 1954) that must be modeled on the finest possible grid without averaging. Thus, the exchange grid , on which this computation is performed, emerges as an independent model component for modeling the surface boundary layer.

The general procedure for solving vertical diffusion is thus split into separate up and down steps. Vertically diffused quantities are partially solved in the atmosphere (known in FMS as the "atmosphere_down" step) and then handed off to the exchange grid, where fluxes are computed. The land or ocean surface models recover the values from the exchange grid and continue the diffusion calculation and return values to the exchange grid. The computation is then completed in the up-sweep of the atmosphere. Note that though we are computing vertical diffusion, some spurious horizontal mixing can occur as the result of regridding.

The exchange grid is described in greater detail in the next section. The complete physical architecture of the FMS coupled system is shown at the end of that section, in Fig. 5.4.

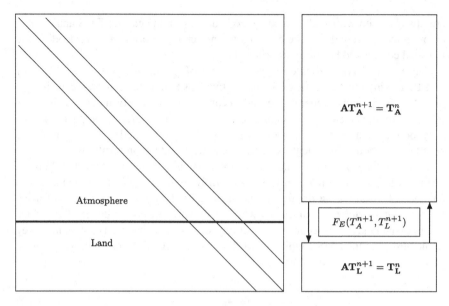

Fig. 5.2 Tridiagonal inversion across multiple components and an exchange grid. The atmospheric and land temperatures T_A and T_L are part of an implicit diffusion equation, coupled by the implicit surface flux on the exchange grid, $F_E(T_A^{n+1}, T_L^{n+1})$

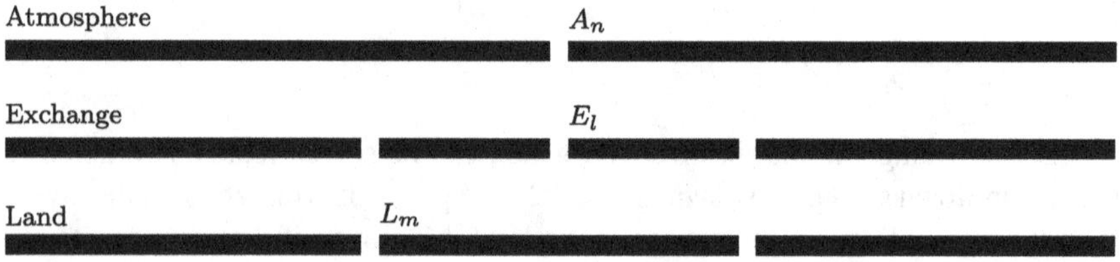

Atmosphere A_n

Exchange E_l

Land L_m

Fig. 5.3 One-dimensional exchange grid

5.4 The Exchange Grid

The exchange grid is described in detail in Balaji et al. (2006).

A *grid* is defined as a set of *cells* created by *edges* joining pairs of *vertices* defined in a discretization. Given two grids, an *exchange grid* is the set of cells defined by the union of all the vertices of the two parent grids. This is illustrated in Fig. 5.3 in 1D, with two parent grids ("atmosphere" and "land"). As seen here, each exchange grid cell can be uniquely associated with exactly one cell on each parent grid, and *fractional areas* with respect to the parent grid cells. Quantities being transferred from one parent grid to the other are first interpolated onto the exchange grid using one set of fractional areas; and then averaged onto the receiving grid using the other set of fractional areas. If a particular moment of the exchanged quantity is required to be conserved, consistent moment-conserving interpolation and averaging functions

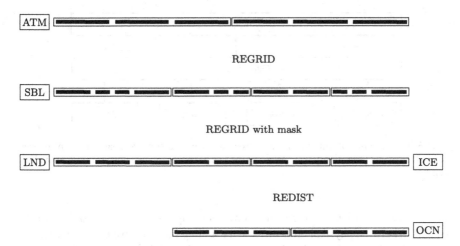

Fig. 5.4 FMS coupled model architecture

of the fractional area may be employed. This may require not only the cell-average of the quantity (zeroth-order moment) but also higher-order moments to be transferred across the exchange grid.

The FMS implementation of exchange grids restricts itself to two-dimensional grids on the planetary surface. However, there is nothing in the exchange grid concept that prevents its use in exchanges between grids varying in 3, or even 4 (including time) dimensions.

Now we consider a further refinement, that of parallelization. In general, not only are the parent grids physically independent, they are also parallelized independently. Thus, for any exchange grid cell E_{nm}, the parent cells A_n and L_m (see Fig. 5.3) may be on different processors. The question arises, to which processor do we assign E_{nm}? The choices are,

1. to inherit the parallel decomposition from one of the parent grids (thereby eliminating communication for one of the data exchanges); or
2. to assign an independent decomposition to the exchange grid, which may provide better load balance.

In the FMS exchange grid design, we have chosen to inherit the decomposition from one side. Performance data (not shown here) indicate that the additional communication and synchronization costs entailed by choosing (2) are quite substantial, and of the same order as the computational cost as the flux computations on the exchange grid itself. Should the computational cost of the exchange grow appreciably, we may revisit this issue.

The complete coupled model architecture is shown in schematic form in Fig. 5.4. The physical components include an atmosphere (ATM), ocean (OCN), land surface (LND), ocean surface (ICE), and surface boundary layer (SBL) component, as described in Sect. 5.3. The schematic depicts the individual grid cells as stripes,

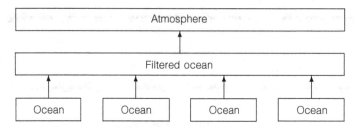

Fig. 5.5 Parallel ensemble filter: four-member ocean ensemble assimilating data and coupled to an atmosphere model. The atmosphere appears to see a single ocean on the same grid as an individual ensemble member

and the boxes surrounding them show how they might be distributed on a parallel system. Where the physical grids differ at a coupling interface, we use regridding by conservative first- or second-order interpolation. Regridding may be under control of a mask, such as at a land-ocean boundary. Where the physical grids are identical (such as between OCN and ICE) but the components may be parallelized differently, a parallel redistribution is performed.

5.5 Data Assimilation

Data assimilation for coupled models is an exciting emergent field of research. Data assimilation includes a class of methods known as *ensemble filters* (Kalnay 2002), which involve sampling the error space of observations by running a model ensemble, i.e. multiple copies of a model perturbed to span that space.

The FMS coupled modeling system includes a sophisticated data assimilation system, the parallel ensemble adjustment Kalman filter (Zhang et al. 2005). The parallel filter consists of running each ensemble member as a concurrent component on an independent set of processors, and the slot replaced by the filter. This is shown schematically in Fig. 5.5, where an ocean-atmosphere model is assimilating ocean data, using a four-member ocean ensemble.

The ensemble filter has been used on 24-member ensembles of CM2.1 with no loss of performance versus ensemble size. A recent development has been the construction of a coupled data assimilation system (CDAS) where ensemble methods simultaneously assimilate both atmosphere and ocean data (Zhang et al. 2007). The CDAS has been the basis for a set of pioneering studies showing the influence of initial conditions on simulations of recent climate history on decadal timescales.

5.6 Conclusions and Perspectives

In summary, this chapter has presented a review of the key features of how coupling is performed in the GFDL Flexible Modeling System. A standard coupling interface with slots for atmosphere, land surface, ocean surface, and ocean components is coupled along with a surface boundary layer component on an exchange grid (Balaji et al. 2006). Components live within a single executable, but can be scheduled serially or concurrently with others. The code has been shown to be scalable to $\mathcal{O}(1,000)$ processors, with fast surface processes coupling every atmospheric timestep (typically ~ 15 min) and slow processes coupling every ocean timestep (typically 1 h). Coupling is conservative to up to second-order accuracy.

The FMS superstructure also includes support for data assimilation using ensemble filter methods. The coupled data assimilation system has been run on IPCC-class models assimilating both atmosphere and ocean fields. An ensemble size of up to 24 has been used with no significant loss of scaling.

At the moment of writing, FMS and its coupler have been in active use for over a decade. Its feature list and its performance still place it at the forefront of the field, and the models built upon it are also acknowledged as being at the top of the scientific line. While FMS principally serves its host institution NOAA/GFDL (~ 100 scientists),[2] the breadth of its influence indicates a useful model for the diffusion of technical design ideas: prototyping within a relatively small community, with the best practices then being adopted in community-wide standards.

References

Balaji V, Anderson J, Held I, Winton M, Durachta J, Malyshev S, Stouffer RJ (2006) The exchange grid: a mechanism for data exchange between earth system components on independent grids. In: Deane A, Brenner G, Ecer A, Emerson D, McDonough J, Periaux J, Satofuka N , Tromeur-Dervout D (eds) Parallel computational fluid dynamics: theory and applications.In: Proceedings of the 2005 international conference on parallel computational fluid ynamics, College Park, Elsevier

Kalnay E (2002) Atmospheric modeling, data assimilation and predictability. Cambridge University Press, Cambridge

Monin A, Obukhov A (1954) Basic laws of turbulent mixing in the ground layer of the atmosphere. Tr Geofiz Inst Akad Nauk SSSR 151:163–187

Zhang S, Harrison MJ, Wittenberg AT, Rosati A, Anderson JL, Balaji V (2005) Initialization of an ENSO forecast system using a parallelized ensemble filter. Month Weath Rev 133:3176–3201

Zhang S, Harrison MJ, Rosati T, Wittenberg A (2007) System design and evaluation of coupled ensemble data assimilation for global oceanic climate studies. Month Weath Rev 135:3541–3564

---

[2] FMS-based atmospheric and ocean models serve larger communities: this statement refers to the coupling architecture.

Chapter 6
The Earth System Modeling Framework

Cecelia DeLuca, Gerhard Theurich and V. Balaji

6.1 Introduction

The Earth System Modeling Framework or ESMF[1] is open source software for building modeling components and coupling them together to form weather prediction, climate, coastal, and other applications. ESMF was motivated by the desire to exchange modeling components amongst centers and to reduce costs and effort by sharing codes. The ESMF package is comprised of a superstructure of coupling tools and component wrappers with standard interfaces, and an infrastructure of utilities for common functions, including calendar management, message logging, grid transformations, and data communications. Infrastructure utilities, including a tool for generation of interpolation weights, can be used independently from the superstructure, enabling users to choose which parts of the software suit their application. The project is distinguished by its strong emphasis on community governance and distributed development.

ESMF was originally designed for tightly coupled applications in the climate and weather domains. Tight coupling is exemplified by the data exchanges between the ocean and atmosphere components of a climate model: a large volume of data is exchanged frequently, and computational efficiency is a primary concern. Such applications usually run on a single computer with hundreds or thousands of proces-

---

[1] http://www.earthsystemmodeling.org

C. DeLuca (✉)
NOAA/CIRES, Boulder CO USA
e-mail: cecelia.deluca@noaa.gov

G. Theurich
Science Applications International Corporation, McLean VA USA

V. Balaji
Princeton University, New Jersey, USA
e-mail: balaji@princeton.edu

S. Valcke et al., *Earth System Modelling – Volume 3*,
SpringerBriefs in Earth System Sciences, DOI: 10.1007/978-3-642-23360-9_6,
© The Author(s) 2012

sors, low-latency communications, and a Unix-based operating system. Almost all components in these domains are written in Fortran, with just a few in C or C++.

As the ESMF customer base has grown to include modelers from other disciplines, such as hydrology and space weather, the framework has evolved to support other forms of coupling. For these modelers, ease of configuration, ease of use, and support for heterogeneous coupling may take precedence over performance. Heterogeneity here refers to programming language (Python, Java, etc. in addition to Fortran and C), function (components for analysis, visualization, etc.), grids and algorithms, and operating systems. In response, the ESMF team has introduced support for the Windows platform, and is exploring approaches to language interoperability through web service interfaces. It has introduced more general data structures and several strategies for looser coupling, where components may be in separate executables, or running on different computers.

In this chapter, we describe how coupling fits within the ESMF architecture, and highlight how specific design strategies satisfy user requirements. We then outline how modelers implement ESMF coupling in their applications. Finally, we review alternative coupling strategies that have evolved to suit new communities using ESMF.

6.2 Architectural Overview

The ESMF architecture is based on the concept of components. At its simplest, a software component is a code that has a well-defined calling interface and a coherent function (e.g. Szyperski 2002). Component-based design is a natural fit for climate modeling, since components are ideally suited for the representation of a system comprised of a set of substantial, distinct and interacting domains, such as atmosphere, land, sea ice and ocean. Further, since Earth system domains are often studied and modeled as collections of sub-processes (radiation and chemistry in an atmosphere, for example) it is convenient to model climate applications as an hierarchy of nested components.

Component-based software is also well-suited for the manner in which climate models are developed and used. The multiple domains and processes in a model are usually developed as separate codes by specialists. The creation of a viable climate application requires the integration, testing and tuning of the pieces, a scientifically and technically formidable task. When each piece is represented as a component with a standard interface and behavior, that integration, at least at the technical level, is more straightforward. Similarly, standard interfaces help to foster interoperability of components, and the use of components in different contexts. This is a primary concern for modelers, since they are motivated to explore and maintain alternative versions of algorithms (such as different implementations of the governing fluid equations of the atmosphere), whole physical domains (such as oceans), parameterizations (such as convection schemes), and configurations (such as standalone versions of physical domains).

There are two types of components in ESMF, Gridded Components and Coupler Components. Gridded Components (ESMF_GridComps) represent the scientific and computational functions in a model, and Coupler Components (ESMF_Cpl Comps) contain the operations necessary to transform and transfer data between them.

Each major physical domain in an ESMF climate or weather model is represented as an ESMF Gridded Component with a standardized calling interface and arguments. Physical processes or computational elements, such as radiative processes or I/O, may also be represented as Gridded Components. ESMF Components can be nested, so that parent components can contain child components with progressively more specialized processes or refined grids.

As a climate or weather model steps forward in time, the physical domains represented by Gridded Components must periodically transfer interfacial fluxes. The operations necessary to couple Gridded Components together may involve data redistribution, spectral or grid transformations, time averaging, and/or unit conversions. In ESMF, Coupler Components encapsulate these interactions.

Design goals for ESMF applications include the ability to use the same Gridded Component in multiple contexts, to swap different implementations of a Gridded Component into an application, and to assemble and extend coupled systems easily; in short, software reuse and interoperability.

A design pattern that addresses these goals is the mediator, in which one object encapsulates how a set of other objects interact (Gamma et al 1995). The mediator serves as an intermediary, and keeps objects from referring to each other explicitly. ESMF Coupler Components follow this pattern. It is an important aspect of the ESMF technical strategy, because it enables the Gridded Components in an application to be deployed in multiple contexts; that is, used in different coupled configurations without changes to their source code. For example, the same atmosphere might in one case be coupled to an ocean in a hurricane prediction model, and in another coupled to a data assimilation system for numerical weather prediction.

Another advantage of the mediator pattern is that it promotes a simplified view of inter-component interactions. The mediator encapsulates all the complexities of data transformation between components. However, this can lead to excessive complexity within the mediator itself, a recognized issue [ibid]. ESMF has addressed this issue by encouraging users to create multiple, simpler Coupler Components and embed them in a predictable fashion in a hierarchical architecture, instead of relying on a single central coupler. This systematic approach is useful for modeling complex, interdependent Earth system processes, since the interpretation of results in a many-component application may rely on a scientist's ability to grasp the flow of interactions system-wide.

Computational environment and throughput requirements motivate a different set of design strategies. ESMF component wrappers must not impose significant overhead, and must operate efficiently on a wide range of computer architectures, including desktop computers and petascale supercomputers. To satisfy these requirements, the ESMF software relies on memory-efficient and highly scalable algorithms (e.g., Devine et al. 2002). Currently ESMF has proven to run efficiently on tens of thousands of processors.

How the components in a modeling application are mapped to computing resources can have a significant impact on performance. Strategies vary for different computer architectures, and ESMF is flexible enough to support multiple approaches. ESMF components can run sequentially (one following the other, on the same computing resources), concurrently (at the same time, on different computing resources), or in combinations of these execution modes. Most ESMF applications run as a single executable, meaning that all components are combined into one program. Starting at a top-level driver, each level of an ESMF application controls the partitioning of its resources and the sequencing of the components of the next lower level.

6.3 Components in ESMF

Both Gridded and Coupler Components are implemented in the Fortran interface as derived types with associated modules. Coupler Components share the same standard interfaces and arguments as Gridded Components. The key data structure in these interfaces is the ESMF_State object, which holds the data to be transferred between components. Each Gridded Component is associated with an import State, containing the data required for it to run, and an export State, containing the data it produces. Coupler Components arrange and execute the transfer of data from the export States of producer Gridded Components into the import States of consumer Gridded Components. The same Gridded Component can be a producer or consumer at different times during model execution.

Modelers most frequently write their own Coupler Component internals using ESMF classes bundled with the framework. These classes include methods for time advancement, data redistribution, calculation of interpolation weights, application of interpolation weights via a sparse matrix multiply, and other common functions. ESMF does not currently offer tools for unit transformations or time averaging operations, so users must manage these operations themselves.

Coupler Components can be written to transform data between a pair of Gridded Components, or a single Coupler Component can couple more than two Gridded Components. Multiple couplers may be included in a single modeling application. This is a natural strategy when the application is structured as an hierarchy of components. Each level in the hierarchy usually has its own set of Coupler Components.

The need for modelers to write their own Gridded or Coupler Components has been changing recently. Generic code is being bundled with ESMF that enables reuse of simple Coupler Components and inheritance from prefabricated Gridded Components. A compliance checker has also been introduced that provides feedback on conformance to conventions during the development process. These additions are making compliance easier and improving interoperability among Components.

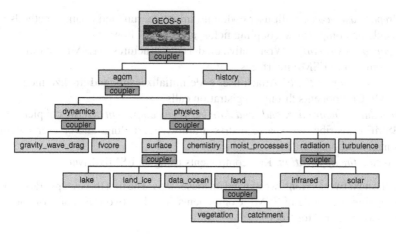

Fig. 6.1 ESMF enables applications such as the GEOS-5 atmospheric general circulation model to be structured hierarchically, and reconfigured and extended easily. Each *box* in this diagram is an ESMF Component

Figure 6.1 shows a simplified schematic of the Goddard Earth Observing System Model, Version 5 (GEOS-5) atmospheric general circulation model, which was constructed in a hierarchical fashion using ESMF. Each box is an ESMF Component. Note that each level in the hierarchy addresses increasingly specific sub-processes represented as Gridded Components, and that each level has its own Coupler Component.

6.4 Remapping in ESMF

Remapping and interpolation in ESMF is accurate, flexible, and fast. ESMF supports a wide variety of grids and remapping options. Generation of interpolation weights and their application is fully parallel. ESMF supports first order conservative, bilinear, and a higher-order finite element-based patch recovery method for remapping in 2D and in some cases 3D. Logically rectangular and unstructured grids are both supported. There is a range of options with respect to masking and the handling of poles and unmapped points. The remapping system is modular; the calculation of interpolation weights can be performed either during a model run or offline, and the application of weights can be made as a separate call.

6.5 Adopting ESMF

It is not necessary to rewrite the internals of model codes to implement coupling using ESMF. Model code attaches to ESMF standard component interfaces via a user-written translation layer that connects native data structures to ESMF data structures. The steps in adopting ESMF are summarized by the acronym *PARSE*:

1. *Prepare user code.* Split user code into initialize, run, and finalize methods and decide on components, coupling fields, and control flow.
2. *Adapt data structures.* Wrap native model data structures in ESMF data structures to conform to ESMF interfaces.
3. *Register user methods.* Attach user code initialize, run, and finalize methods to ESMF Components through registration calls.
4. *Schedule, synchronize, and send data between components.* Write couplers using ESMF redistribution, sparse matrix multiply, regridding, and/or user-specified transformations.
5. *Execute the application.* Run components using an ESMF driver.

In the next two sections, we expand on these steps. The first three steps (*PAR*) focus on wrapping user code in ESMF Components. The last two (*SE*) concern coupling ESMF Components together.

6.5.1 Wrapping User Code in ESMF Components

The first step, *preparing user code*, involves deciding what elements of the application will become Gridded Components. At this time, many climate and weather modeling groups wrap major physical domains (land, ocean, etc.) as Gridded Components, and expect to wrap atmospheric physics and dynamics as Gridded Components in the future. A few applications, such as the GEOS-5 model at the National Aeronautics and Space Administration (NASA), wrap sub-processes such as radiation as Gridded Components as well. A key consideration is what elements of the model are expected to be exchanged or used in multiple contexts; these elements are good candidates for component interfaces. Once Gridded Components are identified, the user must split each of them cleanly into initialize, run, and finalize sections, each callable as a subroutine. These subroutines can have multiple phases; for example, run part one and run part two.

This step also involves analyzing the data flow between components: what fields need to be transferred, what transformations are required between components, how frequently fields must be transferred, and what the data dependencies are. This analysis should give the user a good idea of what Coupler Components will be required, and what operations they should contain. In general, this first step takes the longest.

6.5.2 Adapting Data Structures

The next step in the *PARSE* sequence is *adapting native data structures* to ESMF. Here native model data is copied by reference or value into an ESMF data type. There are multiple ESMF data types that can be used, ranging from simple ESMF_Array objects to ESMF_Field objects that store coordinate information and metadata. Many applications contain multiple physical fields that share the same physical domain. Collections of ESMF arrays and fields can be represented in a compact way using ESMF_ArrayBundle and ESMF_FieldBundle objects.

All data exchanged between components is stored in ESMF import and export State objects. These are simple containers that hold ESMF arrays, arraybundles, fields, and fieldBundles. Once native data structures have been associated with ESMF data types, they must be added by the user to the appropriate State objects. At this point the user code is quite close to the required ESMF interfaces. A remaining task is to wrap native calendar and time information into an ESMF_Clock object. The Clock holds information about start time, stop time, time step, and calendar type, and enables the user to set alarms related to specific events. The modeler may also choose to use the ESMF_Config object to store configuration parameters. Config is a straightforward utility that enables the application to read labels and values from a text file.

The resulting user component methods, for initialize, run, and finalize, have ESMF data structures at the calling interface, and look like this example initialize subroutine:

```
subroutine myOcean_Init(gridComp, importState, &
                        exportState, clock, rc)
   type(ESMF_GridComp)  :: gridComp
   type(ESMF_State)     :: importState
   type(ESMF_State)     :: exportState
   type(ESMF_Clock)     :: clock
   integer, intent(out) :: rc

   ! Wrapping layer in which native arrays are extracted
   ! from model data structures, and referenced or copied
   ! into ESMF Arrays, ArrayBundles, Fields, or
   ! FieldBundles. References to these objects are then
   ! placed into import and export States.
   ! Scientific content of initialize routine goes here.

   rc = ESMF_SUCCESS

end subroutine myOcean_Init
```

6.5.3 Registering User Methods

The third step, *registering user methods*, is relatively simple. In it the user-written part of a Gridded or Coupler Component is associated with an ESMF_GridComp or ESMF_CplComp derived type through a special SetServices routine. This is a routine that the user must write, and declare public.

Inside the SetServices routine the user calls ESMF SetEntryPoint methods that associate the standard initialize/run/finalize ESMF Component methods with the names of their corresponding user code subroutines. For example, a user routine called myOcean_Init might be associated with the standard initialize routine for a Gridded Component named myOcean. The sequence of calls is outlined below.

First the Gridded Component is created. This happens one level above the Gridded Component code. This level may be a relatively small driver program, or it may be a parent Gridded Component. The highest level of an hierarchical ESMF application can be thought of as the "cap". Templates and examples are provided within ESMF to show how the driver is structured.

The application driver would contain code similar to this:

```
type(ESMF_GridComp) :: oceanComp
oceanComp = ESMF_GridCompCreate(name="myOcean", rc=rc)

call ESMF_GridCompSetServices(gridcomp=oceanComp, &
   userRoutine=mySetServices, rc=rc)
```

Here mySetServices is the user given name of the public component routine that is responsible for setting the initialize, run and finalize entry points for ocean Comp. If the Fortran subroutine names of the user's initialize, run, and finalize methods were myOcean_Init, myOcean_Run, and myOcean_Final, respectively, the mySetServices method would contain the following calls:

```
call ESMF_GridCompSetEntryPoint(gridcomp=oceanComp, &
   method=ESMF_SETINIT, userRoutine=myOcean_Init, rc=rc)

call ESMF_GridCompSetEntryPoint(gridcomp=oceanComp, &
   method=ESMF_SETRUN, userRoutine=myOcean_Run, rc=rc)

call ESMF_GridCompSetEntryPoint(gridcomp=oceanComp, &
   method=ESMF_SETFINAL, userRoutine=myOcean_Final, &
   rc=rc)
```

These calls link the two pieces of the component: the Gridded Component derived type provided by the framework and the methods provided by the user. The result is that myOcean model can be dispatched by a driver or by a parent component in a generic way. The create and destroy operations for components are not linked to user code; they act only on the component derived type.

Like the ESMF_GridCompCreate() and ESMF_GridCompSetServi ces() calls, the initialize, run, and finalize methods are invoked from a driver or parent component using standard ESMF-defined Component methods. They would follow the ESMF_GridCompSetServices() call shown previously:

```
call ESMF_GridCompInitialize(gridcomp=oceanComp, ...,
  rc=rc)
call ESMF_GridCompRun(gridcomp=oceanComp, ..., rc=rc)
call ESMF_GridCompFinalize(grdicomp=oceanComp, ...,
  rc=rc)
```

The omitted arguments indicated by "..." are the optional importState, exportState, and clock arguments. The State arguments are necessary to import and export data to and from the component. The Clock argument provides a means to synchronize the simulation time between different model components.

6.5.4 Coupling Between ESMF Components

A very simple ESMF coupled application might involve an application driver cap, a parent Gridded Component, two child Gridded Components (e.g. an oceanComp and an atmComp) that require an inter-component data exchange, and two Coupler Components. The hierarchical structure results in calls cascading so that when, for example, the initialize routine of a parent component is called, it contains and calls the initialize routines of its children, which contain and call the initialize routines of their children, and so on. The result is that a call to a Gridded Component initialize method at the top of the hierarchy initializes all the Components in the hierarchy.

The next step, following the *PARSE* approach, involves *scheduling, synchronizing, and sending* data between Components. A sequence similar to that shown for the Gridded Component oceanComp would be followed in order to create and register methods for a second Gridded Component atmComp and two Coupler Components, oceanToAtmCpl and atmToOceanCpl.

Assuming atmComp needs the temperature field produced by oceanComp, the oceanToAtmCpl Coupler Component is responsible for the correct data flow. If both Gridded Components define the temperature field on the same physical grid, but with their own custom distribution, a simple field redistribution can be used. Otherwise, if the physical grids are different, an interpolation is necessary.

The required pre-computations for coupling are typically carried out during the Coupler's initialize phase, storing the complete exchange pattern in an ESMF_Route Handle object:

```
type(ESMF_RouteHandle) :: routehandle
ESMF_FieldRedistStore(srcField=oceanTempField, &
  dstField=atmTempField, routehandle=routehandle,
  rc=rc)
```

Here the oceanTempField and atmTempField are Fields from the ocean Component's export State and the atmosphere Component's import State, respectively. The actual data exchange between these Field objects takes place during the

Coupler's run phase. A data redistribution call would look like this:

```
call ESMF_FieldRedist(srcField=oceanTempField, &
    dstField=atmTempField, routehandle=routehandle,
    rc=rc)
```

6.5.5 Executing the Application

The last step, *execution*, combines all the pieces into a complete ESMF application. The sequencing is specified in the application driver or parent Component. In the following example, the ocean is run first; the ocean to atmosphere coupler communicates the ocean export State to the atmosphere import State; the atmosphere runs; the atmosphere to ocean coupler communicates the atmosphere export State to the ocean import State. This loop repeats until the stop time of the Clock is reached.

```
do while (.not. ESMF_ClockIsStopTime(clock=clock,
    rc=rc))
    call ESMF_GridCompRun(gridcomp=oceanComp,          &
        importState=oceanImportState,                  &
        exportState=oceanExportState, clock=clock, rc=rc)
    call ESMF_CplCompRun(cplcomp=oceanToAtmCpl,        &
        importState=oceanExportState,                  &
        exportState=atmImportState, clock=clock, rc=rc)
    call ESMF_GridCompRun(gridcomp=atmComp,            &
        importState=atmImportState,                    &
        exportState=atmExportState, clock=clock, rc=rc)
    call ESMF_CplCompRun(cplcomp=atmToOceanCpl,        &
        importState=atmExportState,                    &
        exportState=oceanImportState, clock=clock, rc=rc)
    call ESMF_ClockAdvance(clock=clock, rc=rc)
end do
```

6.6 Alternative Forms of Coupling

Two alternative modes of coupling that have been recently introduced are coupling multiple executables and "direct" coupling. In order to couple multiple executables—wholly separate programs—ESMF has been collaborating with the InterComm project (Lee and Sussman 2004). Through an extension of ESMF Array class methods, users can translate ESMF Arrays into their InterComm equivalent, and perform sends and receives of data to InterComm applications. This work is supporting a space weather application in which an ESMF atmosphere couples to an InterComm ionosphere.

Direct coupling was introduced as a way to initiate a data exchange without needing to first return to a Coupler Component interface. The data exchange is arranged within a Coupler Component, usually at initialization time, but it can be invoked from deep within a Gridded Component. This is useful for many modeling situations, including tightly linked physical processes and asynchronous I/O.

6.7 Conclusions and Perspective

ESMF has successfully built a diverse customer base that includes most of the major climate and weather models in the US. Now in its third funding cycle, ESMF is supported by the National Science Foundation (NFS), the Department of Defense, National Oceanic and Atmospheric Administration (NOAA), and NASA. The development team for ESMF and related projects is maintained at about 12 people. It is based at the NOAA Earth System Research Laboratory and the Cooperative Institute for Research in Environmental Science at the University of Colorado. In addition to adding new features, the team provides dedicated user support, nightly regression testing with a suite that includes thousands of tests and examples, and comprehensive documentation.

Timing results for a variety of codes show that the overhead of using ESMF Components is typically negligible (<3% of runtime), and that key operations scale to tens of thousands of processors. Testing performance and exploring new architectures are ongoing activities. Grid remapping and parallel communications are highly scalable and extensible to many new grid types. The framework is very robust and is supported on more than 24 platforms.

ESMF is continuing to add options and optimizations throughout the code. A recent focus has been adding regridding capabilities, including a first order conservative 3D method and higher order conservative method. Looking to the longer term, the ESMF team is exploring the use of metadata to broker and automate coupling services, and to create self-documenting applications. There is also an effort underway to create an option to register ESMF Components for web service interfaces. This will increase accessibility to coupling services for a broader range of scenarios and communities.

References

Devine K, Boman E, Heaphy R, Hendrickson B, Vaughan C (2002) Zoltan: Data management services for parallel dynamic applications. Comput. Sci. Eng. 4(2):90–97

Gamma E, Helm R, Johnson R, Vlissade J (1995) Design patterns: elements of reusable object-oriented software. Addison Wesley, Bostton

Lee J, Sussman A (2004) Efficient communication between parallel programs with intercomm. Technical report CS-TR-4557 and UMIACS-TR-2004-04, Department of Computer Science and UMIACS, University of Maryland

Szyperski C (2002) Component Software. Addison Wesley, Boston

Chapter 7
The Bespoke Framework Generator

Rupert Ford and Graham Riley

7.1 Introduction

The approach to coupling taken by the Bespoke Framework Generator (BFG) differs from the couplers described in the previous chapters of this Volume. Rather than being a coupler in its own right, BFG allows the user to choose the coupling technology, i.e. a specific coupler and/or communications infrastructure, they would like to use for a coupled model run. Given the required information, in the form of metadata, BFG generates bespoke[1] wrapper code which can be compiled and linked with the users' science code and the coupling technology of choice. Regardless of which coupling technology the user chooses for their coupled model run, the scientific code remains unchanged. BFG can, therefore, be thought of as a Meta-coupler.

The BFG approach originated in a collaborative project with the Met Office: the Flexible Unified Model Environment (FLUME) project. The FLUME project's aim was to design and develop a new, more modular, software architecture for the Met Office's unified model code. One aspect of the architecture is the coupling system.

As part of the FLUME project, the authors performed a requirements analysis for a coupling system for the Met Office. Following the analysis, the properties that a coupler would need to have in order to meet the requirements were identified and a number of existing couplers were evaluated against the properties. The requirements were extracted from meetings with model and infrastructure developers at the Met Office and from discussions with developers at the National Center for Atmospheric Research (NCAR), the European Centre for Medium-Range Weather Forecasts (ECMWF), the Institut Pierre Simon Laplace (IPSL), the Max Planck Institute for Meteorology, the Centre Européen de Recherche et de Formation Avancée en

---

[1] Apparently the word 'bespoke' is not well known outside of the UK. It stands for 'one off', or 'tailored'.

R. Ford · G. Riley (✉)
University of Manchester, Manchester UK
e-mail: graham.riley@manchester.ac.uk

S. Valcke et al., *Earth System Modelling – Volume 3*,
SpringerBriefs in Earth System Sciences, DOI: 10.1007/978-3-642-23360-9_7,
© The Author(s) 2012

Calcul Scientifique (CERFACS) and the National Center for Atmospheric Science (NCAS).

Space constraints preclude a detailed discussion of the results of the requirements analysis; further information can be found in (Ford and Riley, 2002, 2003). It was found that no existing coupling system had all the required properties. This led to a decision to design and prototype a system that did.

The remainder of this chapter describes BFG in more detail and, where appropriate, relates it to the other couplers presented in this Volume and to the Framework descriptions discussed in Volume 1 of this series. We begin by presenting an overview of the BFG approach in Sect. 7.2, introducing the flexibility that this approach provides. The initial version of BFG that was developed (BFG1) is discussed in Sect. 7.3 and the current version, BFG2, is described in Sect. 7.4. This is followed by a review of related systems in Sect. 7.5. The final (Sect. 7.6) presents a summary and discusses possible future developments.

7.2 Architectural Overview

First, a general methodology, termed the Flexible Coupling Approach (FCA), was developed by Ford et al. (2006)[2]. FCA follows a domain specific Generative Programming approach (developing programs that synthesize other programs) and prescribes a class of coupling systems that have the following elements:

• component model implementation rules;
• a language for describing coupled models;
• a framework generator.

The idea is that users (scientists/coupled model developers) describe configurations of coupled models in the language of an FCA-type coupling system and this information is used as input to a framework generator. The framework generator produces, as output, the control and communication source code required, in addition to the FCA-compliant (scientific) source code of the component models involved, to implement the described coupled model. The output of the framework generator is the source code of a bespoke framework along with, where appropriate, configuration files required by any developer-specified existing coupling technology (e.g. OASIS4, see Chap. 4). The source code may then be compiled and linked ready for execution of the coupled model.

By separating the implementation of the coupling technology from the science code the user is given an additional layer of flexibility. This flexibility can help in terms of portability, performance, maintenance and future-proofing of the code. For example, for development purposes a coupled model might be configured to run on a laptop or small desktop machine as a single executable, where coupling data may be exchanged via simple argument passing; for production, the model may be required

---

[2]   This paper uses an earlier acronym for this approach: GCF

to be configured to run (some) models concurrently on a large supercomputer using OASIS to exchange coupling data. With BFG-compliant models, these changes in configuration can be achieved simply with a change of the metadata describing the coupled model and a re-generation of the 'wrapper' code.

A natural thought would be that the high abstraction of a Meta-coupler would necessarily incur an associated run-time performance penalty. However, with BFG this is not the case. By providing a low-level 'scientific' API for use in the science code (described later in this chapter) and by generating bespoke code for a particular coupling implementation and coupled model, BFG can achieve the same performance as hand written code (Armstrong et al. 2009).

The next two sections describe BFG1 and BFG2, two generations of implementations of our FCA-type coupling system.

7.3 BFG1

The Bespoke Framework Generator version 1 (BFG1) (Ford et al. 2006) is a prototype system which was developed as a test bed to determine whether it is possible to build an FCA-type coupling system which can generate wrapper code for a wide range of frameworks (*targets*) suitable for execution on diverse computing resources, ranging from laptops to supercomputers and from tightly coupled parallel systems to distributed computing resources. In this respect it has been successful, as BFG1 can generate wrapper code which uses shared memory buffers, MPI, TDT (see Chap. 2), OASIS3 (see Chap. 4), Globus MPI, or Web Services.

BFG1 specifies a small number of coding rules to which a component model must adhere to be compliant. A component model must be implemented as a single subroutine with no arguments. All coupling input and output for a component model is implemented via in-place calls of the form get(data, tag) and put(data, tag), where data is a reference to the data that is to be input or output and tag is an integer identifier that is unique for all get calls within a component model and is unique for all put calls within a component model.

BFG1 defines XML schemas (Bray et al. 2000) (Thompson et al. 2004) to capture metadata (in XML documents). The metadata documents are structured to reflect a separation of the following concerns from the perspective of a user: **D**escription of individual component models, the **C**omposition of component models into a coupled model, and the **D**eployment of the coupled model onto the selected computing resources. This metadata structure is referred to as the **DCD** (Describe, Compose, Deploy) structure.

The set of Description XML documents describe a compliant component model. These documents use the value of the tag (in a put or get statement) as a key to specify what data a model expects to receive or provide to another model (this information includes the data's type and number of dimensions etc.) and the model's timestep. These documents only need to change when the implementation of a component model changes.

A further XML document captures the details of the Composition of several individual models into a coupled model. A composition consists of:

- a set of point-to-point connections between model component outputs and inputs,
- the sequence of the component models in the coupled model,
- the duration of the run.

Once the composition document describing the coupling between models is complete the science that a coupled model performs is fully defined. Any changes required to the coupling between component models therefore only requires modifications to this document. No changes are required for the component model metadata or the implementation (i.e. the code base for) the component model.

A final metadata document describes how the coupled model is to be Deployed using a particular coupling framework and how the components in the coupled model are to be mapped onto a specified set of executables. Any change in the choice of the target framework (e.g. from OASIS3 to Web Services) is, therefore, limited to a minor modification of the metadata in this Deployment document. When changing the target framework, no changes are required to the composition metadata, the component model metadata and the implementation of the component model, as the science being implemented is the same.

A final 'top level' XML document is used by BFG1 to reference all the XML documents involved in the description of a coupled model.

The BFG1 code generator is implemented Extensible Stylesheet Language Transformations (XSLT, see Clark 1999). Code generation is split into the generation of a control layer for the coupled model and a communications layer based on the specific target framework selected. The control layer XSLT is responsible for calling the component models at the required rates, and for initialising and finalising an underlying communication infrastructure. The control layer is written in a communication framework-independent way and therefore only needs to be written once for the coupled model. The control layer does not require separate generation if a new target framework is selected (through a change to the Deployment metadata). In effect, there is a separate XSLT generator for the communications layer for each possible target. Both the control layer and communications layer XSLT take the top level XML document as input and produce source code as output. For some targets, such as OASIS3, TDT and Web Services, additional XSLT is called to produce the configuration files required by these frameworks.

A PERL script is provided to invoke the XSLT code. The user runs this script with the appropriate 'top level' XML document as an argument. BFG1 then generates the appropriate bespoke wrapper code which the user can compile and link with the component code to create the required executables.

Despite its relatively simple API, BFG1 has proved useful in an Integrated Assessment Framework, called CIAS, which has been developed by the Tyndall Centre in the UK (Warren et al. 2008). This framework links climate component models with impact component models to help investigate mitigation policies to cope with Climate Change.

In a collaboration related to CIAS, with colleagues at PIK in Germany, the possibility of coupling component models written to conform to different frameworks was investigated. In this work, TDT-compliant component models were coupled to BFG1-compliant component models using BFG1. TDT is a socket-based communication infrastructure developed by PIK (see Chap. 2 for more details). The strategy was to use TDT as the communications mechanism and to avoid generating wrapper code for the existing TDT-compliant component models. Whilst this strategy was successful, it was quite limited in scope and many issues remain. Nevertheless, these experiments provided positive feedback for the FCA approach, as implemented in BFG1.

7.4 BFG2

The Bespoke Framework Generator, version 2 (BFG2) (Armstrong et al., 2009) is also a prototype system. At the time of writing, development of this system is ongoing and, to date, has been driven by the desire to demonstrate the ability of an FCA-type system to be able to address the needs of three groups: first, the Met Office, whose requirements were captured in the FLUME project (see Volume 1 of this series), secondly, the Grid ENabled Integrated Earth system model (GENIE) community (http://www.genie.soton.ac.uk/GENIEfy/ also described in Volume 1 of this series) and thirdly, the CIAS Integrated Assessment community (mentioned in the previous section). The combined requirements of these three groups are proving to be a good test of the claimed flexibility of our approach.

BFG2, like its predecessor BFG1, uses an XML-based input language and XSLT processing engine. As in BFG1 users supply XML metadata in three document formats reflecting the DCD (Describe, Compose, Deploy) separation of concerns described earlier. The component model implementation rules for BFG2 are:

- models must be written in Fortran 77 or Fortran 90;
- a model is a collection of subroutines;
- models may communicate data by argument-passing or by calling user-specified send and receive routines in model code (for example, if the send and receive routines are called `put` and `get`, as in BFG1, then communication is of the form `call put(data,tag)` and `call get(data,tag)`, where data is a reference to the data to be communicated, and tag is an integer identification tag). Send routines have non-blocking semantics.
- models must not share data (for example, via common blocks or through use association in FORTRAN90); all inter-model communication must be through the above-mentioned communication mechanisms.

BFG2 therefore extends the BFG1 API to include support for coupling using argument passing and support for component models with a number of subroutines (termed 'entry points'). BFG2 also includes support for the initialisation of component models and arbitrary nesting of loops in the control code ('sequencing') definition, features that were lacking in BFG1. Note, BFG1 models are BFG2-compliant and all that is required to use BFG1 models with BFG2 is to transform the BFG1 metadata into BFG2 metadata. An XSLT transform utility is provided with the BFG2 distribution to do this.

7.4.1 Argument Passing

The ability for component models to be able to input and output data via argument passing (as well as in-place `put` and `get`-style calls) was added to BFG2 for two main reasons: first, to support the way many component models are written and, secondly, to support efficient implementations.

When one analyses the Met Office's Unified Model, many of its component models (which are referred to as sub-models by the Met Office) are written to pass their shared data in and out via arguments. This is also the case for all the GENIE component models (their coding rules require this). Other systems, such as FMS (see Chap. 5), also pass data between component models via arguments, (although in their case the data is defined in structures). This approach is not surprising as argument passing is the natural way to share data when the component models are called in sequence within a single program (i.e. where the subroutine entry point of one component model subroutine is called after that of another model). Some form of message passing is the natural communication mechanism when component models are contained in separate processes where there may be concurrent execution of components.

Argument passing is also the most efficient way to communicate data between component models when they are run in-sequence with each-other. It is efficient in terms of execution time since no expensive copying of data is required to pass the data, and it is efficient in terms of memory as only a single copy of the data needs to be allocated.

When BFG2 component models are coupled together to run in sequence and their coupling consists of data that is passed by argument then BFG2 will generate wrapper code that is equivalent to hand written code (creating a single shared location for the data). If any of the models passes data by in-place `put` or `get` calls then BFG2 will generate code to copy their data into the shared location.

When BFG2 component models are coupled together concurrently then BFG2 will generate the appropriate communications code, using the specified target framework. In the case of data being passed by argument, BFG2 will generate an internal `put` or `get` in the wrapper code to enable this communication.

A natural question at this point might be: Why support in-place `put` and `get` calls if argument passing is the most natural? Argument passing is certainly a natural approach for coupling data, however if one considers the case of a diagnostic output that is calculated deep down in the subroutine hierarchy of a component model, then having to pass this diagnostic all the way up through the argument lists is not necessarily a good idea. Arguably, a better approach is to use in-place `put` calls in this case.

Another question one might ask is: Why does BFG2 not support other forms of data sharing, for example, global memory, modules/common and direct file input and output? The answer to this is that BFG2 could be extended to support these mechanisms but we decided that it was unclear that such support would be used in practise, as there was no strong requirement from any of the application groups with which we were involved.

7.4.2 Subroutine Entry Points

BFG2 supports component models consisting of one or more subroutines, termed entry points. The Fortran90-style API is implemented as a single module which contains these entry points which are declared as public subroutines. In the metadata, subroutines are associated with *initialisation, timestepping* or *finalise* phases of the model, although in practise this separation is not necessary. An arbitrary number of subroutines may be associated with each phase. BFG2 allows the user to specify any ordering constraints between subroutines. For example, a component model may have two separate initialisation routines and one of these may have to run before the other for the component model to run correctly. Normally, all initialisation phase subroutines would run before any timestepping routines.

Therefore, component models, which are frequently, and naturally, written as a number of distinct subroutines, need little modification to be supported by BFG2. The ESMF framework (see Chap. 6) provides a similar level of support for multiple entry points models.

7.4.3 Scientific API

The core concept of the BFG2 component model API is to support the way modellers write their scientific code. The idea is to *isolate* the science that a component model performs—effectively defining a *scientific API* for the model—from the code which implements control over the execution of models in a coupled model and from the code which enables the communication of coupling data between models.

As models are generally written and maintained by scientists, the only intelligence that we suggest is coded into a component model is that concerned with the science itself. In BFG2, a component model must know how to compute a single timestep of the integration of the model's equations, and the data it needs from other models to do so. The model also needs to know about the data that it can provide to other models. The data required and provided must be specified in metadata documents. The philosophy is to capture non-scientific information as metadata rather than code it explicitly into the component model. This is the source of BFG's flexibility. Having an essentially scientific-oriented component model ensures that the model is free from framework bias and from the details of any particular programming methodology.

7.4.4 Initialisation

BFG2 supports the initialisation of a component model's data through various means. Initialisation is called *priming* in BFG2 as one of the main tasks of initialisation is to provide data to some of the coupling connections the first time a component model is called. This is needed in order to start the model off. Once running, the coupling data will generally be provided by other component models. In BFG2, priming data can be:

- provided by data from another entry point, which may be an entry point of the same or a different component model (and specified in the composition metadata);
- provided internally by the entry point (e.g. as a constant coded by the developer);
- provided by a predefined value specified in the metadata;
- provided from a file, either in namelist or NetCDF format.

BFG2 processing will generate the appropriate source code in each case.

7.4.5 Control

BFG2 supports the specification of component models that are called from within an arbitrary nest of control loops. BFG2 does not currently support conditional execution of component models. This level of support has been found to be sufficient for all known configurations of coupled models emerging from our collaborations with the Met Office, GENIE and CIAS communities. Anticipated future developments for the control metadata include providing support for component models that are required to be iterated to convergence, for example, under the control of a 'do while' loop, and for models for which control is event-based rather then timestepping-based.

7.4.6 XSLT Implementation

The complexity of the BFG2 metadata specification meant that it was not feasible to translate the metadata directly into code in a single transformation phase. In particular, the need to support coupling via argument passing added a significant amount of complexity when compared with BFG1.

The solution taken to this problem was to split the code generation into a number of phases. Code generation begins with a generic code template (which is internal to BFG2) and this is gradually turned into the required code in a series of calls to XSLT templates. This approach appears to be a good one in theory, however in practise (at least in our implementation) one or two phases have become very complex. It is possible that these complex phases could be split into further, simpler phases.

An advantage of using a code template that is gradually expanded is that this template can maintain information (state) from one template call to the next. Those who use XSLT will know that it is a single assignment functional language and, therefore, does not naturally maintain state.

7.4.7 Frameworks as Targets

As already discussed, BFG2 supports coupling via argument passing as well as via in-place calls, with any combination of these two mechanisms allowed. Coupling via argument passing can be used when the data (variables) being coupled are subroutine arguments and the component models will be deployed within a single executable (where, typically, the models will run 'in sequence'). If this is not the case (i.e. either the coupling is via in-place calls or the models will be deployed in multiple executables) then, currently, the coupling between variables must be implemented either using MPI or OASIS4, depending on which is specified in the metadata. Work is underway to extend the list of targets supported in BFG2. In particular, we are adding support for ESMF.

7.4.8 Grids

Spatial grids are widely used in Earth System Models. A component model may use several types of grids, each associated with a specific data field, and the grids used by other component models in a coupled model may well be different again. Grids typically vary either in how the grid points are defined—for example, there are both regular and irregular grids—or simply in the spatial resolution of their grid points.

One of the perceived benefits of coupling systems is that they can provide facilities to transform data from one type of grid to another. An on-going international effort, *Gridspec*[3], aims to support grid definitions and grid transformations in a standard way. A prototype XML implementation of the Gridspec has now been defined and support for this prototype Gridspec has been added to BFG2 in order to demonstrate that BFG2 is able to support grids. The Gridspec support is used to generate appropriate OASIS4 grid transformation metadata, where it is required, and also to support the definition of coded (i.e. hand written) grid transformations.

---

[3]   http://www.gfdl.noaa.gov/~vb/gridstd/gridstd.html

7.4.9 The GENIE ESM—An Example

The GENIE ESM (see Volume 1 of this series) has been used as a test-bed for BFG2. In the original implementation of GENIE, the GENIE component models were implemented as argument passing subroutines. The coupling between component models was implemented in a hand written 'main' code in which component models are called in a pre-specified sequence, resulting in a coupling (composition) of models which is implied by the argument passing that occurs as the code executes.

The hand written main code contained all the possible combinations of component models, and an input configuration file defined the actual component models to be used in a particular run of the ESM.

A number of the GENIE component models have been made BFG-compliant. This did not take much effort, as almost no code changes were required to the code of the component models. Specifying the composition metadata, however, took a lot longer. This task involved making the implicit composition encoded in the hand written main code explicit. Once this task was completed, BFG2 was able to reproduce the original functionality of certain configurations of the ESM which ran with *the same performance* as the original hand crafted code. In fact, BFG2 uses slightly less memory than the equivalent hand crafted version (since the original version declared the data for all possible configurations of models and BFG2 only declares the data needed for the specific composition of models specified in the metadata).

Not only can BFG2 generate argument passing code that has equivalent performance to the hand crafted code for the same composition of models, but by modifying the metadata appropriately one can then choose different GENIE component model orderings (that are not possible in the existing hand written code). Further, with another minor change to the deployment metadata, it is possible to choose to run component models concurrently, if required. As an example, one particular concurrent configuration of GENIE results in a performance improvement of almost a factor of two on a dual core processor (Armstrong et al., 2009).

7.5 Related Systems

BFG is essentially a coupling framework *generator* rather than a coupling framework. Thus, BFG can usefully be considered to be a Meta-Coupler. BFG's API is designed to be sufficiently flexible, and its metadata is designed to be sufficiently abstract, to allow it to generate wrapper code for existing coupling frameworks such as OASIS3 or OASIS4 (see Chap. 4), ESMF (see Chap. 6) and TDT (see Chap. 2). However BFG can also create wrapper code for an MPI-based implementation of coupling and couple component models via argument passing etc. When used in this latter way, BFG actually behaves as a coupler itself. The distinction between a framework generator and a framework is obviously not a completely clean one.

The proposed approach to coupling in the FLUME project at the Met Office, unsurprisingly, follows a very similar approach to BFG. The FLUME API is very similar to the BFG API, so FLUME-compliant models will also be BFG-compliant. However, the Met Office has taken a different approach to wrapper code generation. Their approach attempts to reduce the complexity of code generation by making the framework code generic wherever possible. In contrast BFG has no generic code. At the present time it is unclear if one of these approaches is generally better than the other.

Apart from FLUME, the system that is most similar to BFG, in terms of the use of code generation, is PALM[4] and the origins of the BFG approach can be traced back to an evaluation of PALM by the FLUME project. This evaluation was undertaken by the authors acting as consultants on the FLUME project. However, PALM limits its code generation to control code, using MPI as the underlying communications layer. Therefore, PALM does not follow the framework generation approach of BFG. An additional technical difference is that the latest version of BFG also supports argument passing as well as in-place calls. PALM supports only in-place calls.

BFG implements the same 'sandwich' approach described for FMS in Chap. 5. Such an approach allows the maximum flexibility in composition and deployment. However, it has been argued that this approach requires a much greater amount of code modification for existing ESM models than the, arguably less invasive but less flexible, library-based approach of systems such as OASIS3, OASIS4 and TDT. Whilst this is certainly the case for codes implemented in non-modular fashion, and particularly for codes which make extensive use of common blocks, it is arguably not the case for more modular systems such as GENIE and FMS.

7.6 Conclusions and Perspectives

The BFG Meta-Coupler family of framework generators uses a combination of Generative Programming (i.e. code generation) based on metadata input and an intuitive 'scientific' API. In particular, the BFG2 API is unique in allowing efficient coupling via argument passing to be achieved. Whilst being prototypes, the two versions of BFG promise the ability to produce both flexible and efficient coupler code. BFG1 is able to produce coupling code for a number of frameworks, including OASIS3 and TDT. BFG2 is able to produce coupling code for OASIS4 and also provides some support for the generation of the coupling code needed when spatial grids are used. The approach has been tested using the GENIE (see Volume 1 of this series) and CIAS (Warren et al., 2008) systems and is influencing developments in the FLUME project.

There are a number of areas where further development and improvement of BFG are required. Some of these are listed below.

---

[4] http://www.cerfacs.fr/globc/PALM_WEB/index.html

- **Supporting more than one framework in a coupled model.** At the moment only one framework can be chosen to implement all the communication for a particular coupled model. For example, one could choose OASIS4 for some of the coupling and MPI for any remaining coupling.
- **Supporting additional targets.** At the moment BFG2 supports OASIS4 and MPI as targets. The plan is to add support for TDT, OASIS3 and ESMF in the near future. Plans are also in place to use MCT for any MxN communication within the MPI target implementation.
- *Support for component models conforming to other framework rules.* Some limited testing of this has been tried with TDT-conforming models; it is however, in general, a difficult problem due to the need to support the full API of each framework.
- *Support for 'minimal compliance' rules.* It would be possible to allow (a number of) component models to be controlled by their own main code and for them to communicate solely via library calls (in a similar way that OASIS3, OASIS4 and TDT are used). The aim would be to integrate such a (set of) model(s) with other BFG-compliant models to produce a complete coupled model. The conformance rule set for these models is termed a 'minimal compliance' rule set, since the rules would be a subset of the full compliance rules. These rules would represent the minimum modifications one would have to make to an existing code for it to be integrated with other BFG-compliant models.
- *Support different types of component model.* FLUME identified component models which were termed *Service models*. For example, a diagnostics component which would service diagnostic requests on behalf of other component models, or an advection component which would calculate advection once and provide the results to a number of other models. Such models can receive coupling data from a number of different component models and may receive the data in any order. Another example of a different type of model is one that supports a variable timestep.
- *Support for event based control.* Future ESMs may contain component models that only exist transiently. As a trivial example, a sea-ice model is only required while sea-ice exists. Hurricane or storm models are other examples of models of transient phenomena.
- *Simplify the Code Generation System.* The code generation XSLT in BFG2 is relatively complex and could be simplified to make the system more maintainable.

At the time of writing, a number of these developments are taking place under EC-funded projects such as METAFOR (http://metaforclimate.eu/, see Volume 1 of this series), IS-ENES[5] and ERMITAGE.[6] It is hoped that, eventually, the best and most useful of BFG features will be absorbed into the coupling infrastructure used by the mainstream Earth System Modelling groups in Europe, the US and in the wider community. In particular, it is anticipated that smaller ESM groups would benefit from the flexibility to use existing community-based models and resources, such as the Grid and web services, effectively.

---

[5] https://is.enes.org/

[6] A recently funded EC project which will couple together a set of European Integrated Assessment Models covering climate change, land use, and the economy.

References

Armstrong CW, Ford RW, Riley GD (2009) Coupling integrated earth system model components with bfg2. Concurrency Computat. Pract. Exper. 21(6):767–791. doi http://dx.doi.org/10.1002/cpe.v21:6

Bray T, Paoli J, Sperberg-McQueen C, Maler E, Yergeau F (2000) Extensible markup language (XML) 1.0. W3C Recommendation 6

Clark J (1999) XSL Transformations (XSLT) 1.0. W3C Recommendation

Ford RW, Riley GD (2002) Model coupling requirements. Flume report, Met Office, http://www.cs.manchester.ac.uk/cnc/projects/bfg.php#papers

Ford RW, Riley GD (2003) Single model software architecture v1.2. Flume report, Met Office, http://www.cs.manchester.ac.uk/cnc/projects/bfg.php#papers

Ford RW, Riley GD, Bane MK, Armstrong CW, Freeman TL (2006) Gcf: a general coupling framework.Concurrency and Computation: practice and experience 18(2):163–181. doi http://dx.doi.org/10.1002/cpe.v18:2

Thompson H, Beech D, Maloney M, Mendelsohn N (2004) XML Schema, W3C Recommendation

Warren R, de la Nava Santos S, Arnell N, Bane M, Barker T, Barton C, Ford R, Fssel HM Hankin, RK, Klein R, Linstead C, Kohler J, Mitchell T, Osborn T, Pan H, Raper S, Riley, G, Schellnhber H, Winne S, Anderson D (2008) Development and illustrative outputs of the community integrated assessment system (cias), a multi-institutional modular integrated assessment approach for modelling climate change. Environ Modelling Softw 23(5):592–610 doi http://dx.doi.org/10.1016/j.envsoft.2007.09.002

Chapter 8
Future Perspectives

Sophie Valcke

This volume presents some of the current coupling technologies used in Earth System Modelling. Basic features shared by all technologies include the ability to communicate and regrid data between the component models. But the existence of different implementations highlights the trade-offs between the different approaches. Direct coupling using a fully concurrent multiple executable approach (such as in OASIS) is somewhat less flexible and in some cases less efficient, but it is relatively straight-forward to implement, requiring only minimal modification to individual component model codes. Coupling via top-level interfaces within one integrated application (such as in ESMF or FMS) requires some standardization around high level interfaces, design, and datatypes, but provides opportunities to run models in more flexible and more efficient configurations. Coupling toolkits (such as MCT or TDT) offer an a la carte use of some specific coupling functions. Finally, research in generative programming (such as the BFG project) explores potential ways to unify the different coupling approaches.

Over the past two decades, the different coupling technology developers have worked relatively independently. As a consequence, the different groups benefit today from fundamentally different solutions, even if these solutions share some common functionality. In the coming years, an effort should be made to encourage the interaction and the sharing of some of the building blocks among the coupling technologies. Best practices in coupling should be discussed, identified and shared.

In the longer term, coupling technologies will have to adapt to future computer platforms likely to consist of orders of magnitude more processors which will be slower, more heterogeneous and with less and slower memory. Therefore, developers will have to be particularly careful about the technological choices they make. New levels of parallelism, in particular an increased concurrency in the components of a coupled system, will have to be found. The localization of the coupling data on the different component processes and the individual component

S. Valcke (✉)
CERFACS, Av. Coriolis 42, 31057 Toulouse Cedex 01, France
e-mail: sophie.valcke@cerfacs.fr

S. Valcke et al., *Earth System Modelling – Volume 3*,
SpringerBriefs in Earth System Sciences, DOI: 10.1007/978-3-642-23360-9_8,
© The Author(s) 2012

decomposition will have to be considered with care in order to minimize data transfer linked to this increased concurrency. New communication strategies such as non-blocking communication overlapping computation will be needed. Given the complexity of the future hardware and high resolution climate systems, unifying the different coupling approaches and therefore combining available resources should be considered to address these upcoming challenges.

Glossary

ACPI	Accelerated Climate Prediction Initiative
ANSI	American National Standards Institute
API	Application Programming Interface
ARPEGE	Action de Recherche Petite Echelle Grande Echelle
ASCII	American Standard Code for Information Interchange
AWI	Alfred Wegener Institute
BFG	Bespoke Framework Generator
BoM	Bureau of Meteorology
CCA	Common Component Architecture
CCSM	Community Climate System Model
CDAS	Coupled Data Assimilation System
CERFACS	Centre de Recherche et de Formation Avancée en Calcul Scientifique
CIAS	Community Integrated Assessment System
CICE	Los Alamos Sea Ice Model
CINES	Centre Informatique National de l'Enseignement Supérieur
CNRS	Centre National de la Recherche Scientifique
CORBA	Common Object Request Broker Architecture
cpu	Computing processor unit

CSIRO	Commonwealth Scientific and Industrial Research Organisation
DKRZ	Deutsches Klimarechenzentrum GmbH
ECMWF	European Centre for Medium range Weather Forecasts
ENEA	Ente Nazionale per le Nuove tecnologie, l'Energia el Ambiente
ERMITAGE	Enhancing Robustness and Model Integration for The Assessment of Global Environmental Change
ESM	Earth System Model
ESMF	Earth System Modeling Framework
EU	European Union
FAQs	Frequently Asked Questions
FCA	Flexible Coupling Approach
FLUME	Flexible Unified Model Environment
FMS	Flexible Modeling System
FOAM	The Fast Ocean Atmosphere Model
GAMS	General Algebraic Modeling System
GEMS	Global and regional Earth-system (Atmosphere) Monitor-ingusing Satellite and in-situ data
GENIE	Grid ENabled Integrated Earth system modelling framework
GEOS-5	Goddard Earth Observing System Model, Version 5
GFDL	Geophysical Fluid Dynamics Laboratory
GPL	GNU Public Licence
GUI	Graphical User Interface
IFM-GEOMAR	Leibniz Institut für Meereswissenschaften an der Universität Kiel
IFS	Integrated Forecast System
INGV	Istituto Nazionale di Geofisica e Vulcanologia
IPSL	Institut Pierre-Simon Laplace
IS-ENES	InfraStructure for the European Network for Earth SystemModelling
JPL	Jet Propulsion Laboratory
KNMI	Koninklijk Nederlands Meteorologisch Instituut
LMD	Laboratoire de Méteorologie Dynamique

LODYC	Laboratoire d'Océanographie Dynamique et de Climatologie
MACC	Monitoring Atmospheric Composition and Climate
MCT	Model Coupling Toolkit
METAFOR	Common Metadata for Climate Modelling Digital Repositories
MOM4	Modular Ocean Model version4
MPEU	Message Passing Environment Utilities
MPI	Message Passing Interface
MPI-M	Max-Planck Institute for Meteorology
MPMD	Multiple Program Multiple Data
NASA	National Aeronautics and Space Administration
NCAR	National Center for Atmospheric Research
NCAS	National Centre for Atmospheric Science
NEMO	Nucleus for European Modelling of the Ocean
NetCDF	Network Common Data Form
NLE-IT	NEC Laboratories Europe-IT Research Division
NOAA	National Oceanic and Atmospheric Administration
NSF	National Science Foundation
OASIS	Ocean Atmosphere Sea Ice Soil coupler
OPA	Océan PArallélisé
PALM	Projet d'Assimilation par Logiciel Multiméthodes
PIK	Potsdam Institut für Klimafolgenforschung
POSIX	Portable Operating System Interface
PRISM	Partnership for Research Infrastructure in Earth System Modeling
PSAS	Physical-space Statistical Analysis System
PSMILe	PRISM System Model Interface Library (OASIS coupler)
PVM	Parallel Virtual Machine
RIVM	Rijksinstituut voor Volksgezondheit en Milieu
SCI	Standard Component Interface
SCRIP	Spherical Coordinate Remapping and Interpolation Package

SMHI	Swedish Meteorological and Hydrological Institute
SVIPC	System V Interprocess Communication
TDT	Typed Data Transfer
UM	Unified Model
XML	Extensible Markup Language
XSLT	Extensible Stylesheet Language Transformations

Index

S. Valcke et al., *Earth System Modelling – Volume 3*,
SpringerBriefs in Earth System Sciences, DOI: 10.1007/978-3-642-23360-9,
© The Author(s) 2012